McDougal Littell

Pre-Algebra

Larson Boswell Kanold Stiff

CHAPTER 3

Resource Book

The Resource Book contains a wide variety of blackline masters available for Chapter 3. The blacklines are organized by lesson. Included are support materials for the teacher as well as practice, activities, applications, and project resources.

A DIVISION OF HOUGHTON MIFFLIN COMPANY
Evanston, Illinois • Boston • Dallas

Contributing Authors

The authors wish to thank **Jessica Pflueger** for her contributions to the Chapter 3 Resource Book.

ISBN: 0-618-26939-8

123456789-QDI-09 08 07 06 05 04

Contents

CHAPTER 3 Multi-Step Equations and Inequalities

Chapter Support	1–4
3.1 Solving Two-Step Equations	5–14
3.2 Solving Equations Having Like Terms and Parentheses	15–23
3.3 Solving Equations with Variables on Both Sides	24–34
3.4 Solving Inequalities Using Addition or Subtraction	35–44
3.5 Solving Inequalities Using Multiplication or Division	45–53
3.6 Solving Multi-Step Inequalities	54–61
Review and Projects	62–70
Resource Book Answers	A1–A9

Contents

CHAPTER SUPPORT MATERIALS	
Tips for New Teachers	p. 1
Parents as Partners	p. 3

LESSON MATERIALS	3.1	3.2	3.3	3.4	3.5	3.6
Lesson Plans (Reg. & Block)	p. 5	p. 15	p. 24	p. 35	p. 45	p. 54
Activity Master		p. 17				
Activity Support Master	p. 7		p. 26			
Tech. Activities & Keystrokes			p. 27	p. 37	p. 47	
Practice A	p. 8	p. 18	p. 29	p. 38	p. 48	p. 56
Practice B	p. 9	p. 19	p. 30	p. 39	p. 49	p. 57
Practice C	p. 10	p. 20	p. 31	p. 40	p. 50	p. 58
Study Guide	p. 11	p. 21	p. 32	p. 41	p. 51	p. 59
Real-World Problem Solving	p. 13			p. 43		
Challenge Practice	p. 14	p. 23	p. 34	p. 44	p. 53	p. 61

REVIEW AND PROJECT MATERIALS	
Chapter Review Games and Activities	p. 62
Real-Life Project with Rubric	p. 63
Cooperative Project with Rubric	p. 65
Independent Project with Rubric	p. 67
Cumulative Practice	p. 69
Resource Book Answers	p. A1

Contents

Descriptions of Resources

This Chapter Resource Book is organized by lessons within the chapter in order to make your planning easier. The following materials are provided:

Tips for New Teachers These teaching notes provide both new and experienced teachers with useful teaching tips for each lesson, including tips about common errors and inclusion.

Parents as Partners This guide helps parents contribute to student success by providing an overview of the chapter along with questions and activities for parents and students to work on together.

Lesson Plans and Lesson Plans for Block Scheduling This planning template helps teachers select the materials they will use to teach each lesson from among the variety of materials available for the lesson. The block-scheduling version provides additional information about pacing.

Activity Support Masters These blackline masters make it easier for students to record their work on selected activities in the Student Edition.

Technology Activities with Keystrokes Keystrokes for various models of calculators are provided for each Technology Activity in the Student Edition where appropriate, along with alternative Technology Activities for selected lessons.

Practice A, B, and C These exercises offer additional practice for the material in each lesson, including application problems. There are three levels of practice for each lesson: A (basic), B (average), and C (advanced).

Study Guide These two pages provide additional instruction, worked-out examples, and practice exercises covering the key concepts and vocabulary in each lesson.

Real-World Problem Solving These exercises offer problem-solving activities for the material in selected lessons in a real world context.

Challenge Practice These exercises offer challenging practice on the mathematics of each lesson for advanced students.

Chapter Review Games and Activities This worksheet offers fun practice at the end of the chapter and provides an alternative way to review the chapter content in preparation for the Chapter Test.

Projects with Rubric These projects allow students to delve more deeply into a problem that applies the mathematics of the chapter. Teacher's notes and a 4-point rubric are included. The projects include a real-life project, a cooperative project, and an independent extra credit project.

Cumulative Practice These practice pages help students maintain skills from the current chapter and preceding chapters.

CHAPTER 3

Tips for New Teachers

For use with Chapter 3

Concept Activity 3.1

TEACHING TIP Be sure to fully develop the concept of zero pairs. Relate zero pairs to the Identity Property of Addition, reminding students that adding or subtracting zero yields the same number. Thus removing zero pairs does not unbalance an equation.

Lesson 3.1

INCLUSION You may wish to have students model some of the Examples and exercises in this lesson using algebra tiles.

TEACHING TIP When solving two-step equations, remind students that they will reverse the order of operations in the solution process. Therefore, they will undo addition or subtraction and then undo multiplication or division.

TEACHING TIP Make sure that students get in the habit of showing each step when solving multi-step equations. Creating this good problem solving technique will pay off as the equations increase in complexity.

TEACHING TIP Continue to emphasize the importance of checking solutions by substituting them back into the original equations.

TEACHING TIP In Example 3 on page 121, some students may need to see an additional step written after step 2:

$$7 - 4y - 7 = 19 - 7$$

Use the commutative property and write:

$$-4y + 7 - 7 = 19 - 7$$

This may help students see the next step

$$-4y = 12$$

more clearly.

Lesson 3.2

TEACHING TIP In Example 1 on page 125, point out to students that the variable, n, is assigned to be both the number of pompoms and the number of noisemakers because they are equal numbers. Thus, solving for n gives both the number of pompoms and the number of noisemakers simultaneously.

TEACHING TIP In Example 3 on page 126 ask students why using the distributive property in step 2 yields:

$$5x - 2x + 2,$$

and not

$$5x - 2x - 2.$$

Make sure students understand that it is -2 that is distributed and thus $(-2)(-1) = 2$.

Concept Activity 3.3

TEACHING TIP If algebra tiles are not available, students may use graph paper and rulers to complete the exercises that involve drawing diagrams.

Lesson 3.3

TEACHING TIP In Example 1 on page 131, some students may prefer to keep the variable on the left hand side of the equation. You may wish to show them the following alternative way to solve this exercise.

$$7n - 5 = 10n + 13$$

$$7n - 5 - 10n = 10n + 13 - 10n$$

(Subtract $10n$ from each side.)

$$-3n - 5 = 13$$

$$-3n - 5 + 5 = 13 + 5 \text{ (Add 5 to each side.)}$$

$$-3n = 18 \text{ (Simplify.)}$$

$$\frac{-3n}{-3} = \frac{18}{-3} \text{ (Divide by } -3.\text{)}$$

$$n = -6 \text{ (Note same solution.)}$$

Tips for New Teachers

For use with Chapter 3

INCLUSION As with Lesson 3.1, you may wish to have students model some of the Examples and exercises in this lesson using algebra tiles.

TEACHING TIP After doing Example 4, you may wish to tell students that an equation that has every number as a solution is called an *identity*.

Lesson 3.4

TEACHING TIP When graphing inequalities, students are often confused when to use an open dot or a closed dot. Have students imagine that the line at the bottom of $\leq$ or $\geq$ is a stick of gum. If a stick of gum is present, they can take it, chew it, and stick it inside the open dot to make it a closed dot. If there's no gum, as in $<$ or $>$, then the dot stays empty and open.

COMMON ERROR The symbol $\leq$ means *less than **or** equal to*, not *less than **and** equal to*. Therefore, it is true that $4 \leq 4$ and also that $4 \leq 5$. However, it is not true that $4 < 4$.

COMMON ERROR In Example 3 on page 139, be sure students understand that $2 > x$ means the same thing as $x < 2$, and *not* $x > 2$. As an example, tell students that if Chris is *older* than Jaime, then Jaime is *younger* than Chris. If the order is switched, than the comparison must be reversed also.

Lesson 3.5

COMMON ERROR Be sure that in addition to remembering when to reverse the direction of an inequality symbol, students also remember when *not* to reverse its direction. Some students may start reversing inequality symbols when it is not appropriate. As an example, have students start with the true inequality $4 < 8$. Ask them to multiply both sides by 2. They will get $8 < 16$, which is still true and thus, it is not necessary to reverse the symbol. Then have them multiply both sides by -2 without reversing the symbol. They will get the inequality $-8 < -16$, which is false. In order to get a true statement, they must reverse the symbol and write $-8 > -16$. Show students similar examples for division, as well.

Lesson 3.6

TEACHING TIP Most students find it easier to graph an inequality when the variable is on the left hand side. Encourage students to rewrite inequalities with a variable on the right hand side to an equivalent inequality with the variable on the left before attempting to graph the solution.

TEACHING TIP Remind students that when solving multi-step inequalities they follow the same process as solving multi-step equations. Students should combine like terms, undo additions and/or subtractions then undo multiplications and/or divisions. It is only in the last step that the sign of the inequality may need to be reversed.

CHAPTER 3

Name ______________________ Date __________

Parents as Partners

For use with Chapter 3

Chapter Overview **One way you can help your student su... by discussing the lesson goals in the chart below. When a les... k your student the following questions. "What were the goals of t... words and formulas did you learn? How can you apply the ideas o... ife?"**

Lesson Title	*Lesson G...*	*Applications*
3.1 Solving Two-Step Equations	Solve two-step equa...	• ...m Set • ... Repair • ...ving • ...ting • ...ins
3.2 Solving Equations Having Like Terms and Parentheses	Solve equations using the distributive property.	• School Spirit • Fishing • Karaoke • Cell Phones
3.3 Solving Equations with Variables on Both Sides	Solve equations with variables on both sides.	• Spanish Club • Shopping • Toll Booth • Pasta Machine
3.4 Solving Inequalities Using Addition or Subtraction	Solve inequalities using addition or subtraction.	• Science • Triathlon • Astronauts • Bacteria • Train Travel
3.5 Solving Inequalities Using Multiplication or Division	Solve inequalities using multiplication or division.	• Geese Migration • Marathon Training • In-Line Skates • Reading • Biking
3.6 Solving Multi-Step Inequalities	Solve multi-step inequalities.	• Soccer • Ice Skating • Amusement Parks • Movie Rental • Advertising

Notetaking Strategies

Summarizing is the strategy featured in Chapter 3 (see page 118). Encourage your student to put in his/her notebook summaries of the main ideas from each lesson. This method is not only an excellent tool to help prepare for end-of-year exams, but also a great step your student can take to integrate and internalize the concepts he/she has learned. This strategy can also provide a framework within your student's mind for remembering the material much more readily in the future.

CHAPTER **3** Continued

Name ______________________ Date ________

Parents as Partners

For use with Chapter 3

Key Ideas Your student can demonstrate understanding of key concepts by working through the following exercises with you.

Lesson	*Exercise*
3.1	Solve the equation. Check your solution. (a) $43 = 13x - 22$ (b) $\frac{m}{7} + 30 = 46$ (c) $21 = 37 - 4t$
3.2	Find the value of x for the given rectangle whose perimeter is 34 units. 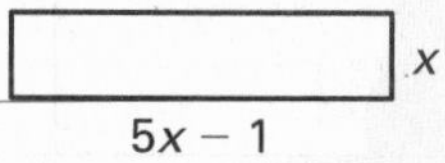
3.3	Solve the equation $57 + 3y = 17 - 7y$. Check your solution.
3.4	Solve the inequality $-9 \geq p - 12$. Graph your solution.
3.5	Solve the inequality $7x > -98$. Graph your solution.
3.6	Solve the inequality $\frac{k}{-4} + 5 \geq 8$. Graph and check your solution.

*Never leave a negative variable

Include area!

Example, *Nega. fractions

Home Involvement Activity

Directions: Write 10 equations on the lined side of 10 index cards. Write the answers to the equations on the lined side of 10 other index cards. Mix up the cards and lay them out on a table with the lined side facing down. Flip over two cards to match the equation with its answer. When you find all the matches, you win!

$10 - \frac{t}{5} > 4$

$t > 30$

Answers

3.1: (a) 5 (b) 112 (c) 4 **3.2:** 3 **3.3:** -4 **3.4:** $p \leq 3$; 

3.5: $x > -14$; 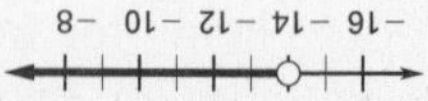 **3.6:** $k \leq -12$; 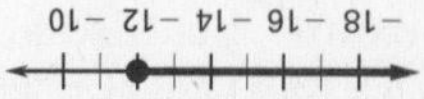

LESSON **3.1**

Teacher's Name ______________________ Class ______ Room ______ Date ______

Lesson Plan

1-day lesson (See *Pacing and Assignment Guide*, TE page 116A)

For use with pages 119–124

GOAL **Solve two-step equations.**

State/Local Objectives ____

____

✓ **Check the items you wish to use for this lesson.**

STARTING OPTIONS

_____ Warm-Up: Transparencies

TEACHING OPTIONS

_____ Notetaking Guide

_____ Concept Activity: SE page 119

_____ Activity Support Master: CRB page 7

_____ Examples: 1–4, SE pages 120–122

_____ Extra Examples: TE pages 121–122

_____ Checkpoint Exercises: 1– 12, SE pages 120–121

_____ Concept Check: TE page 122

_____ Guided Practice Exercises: 1–7, SE page 122

APPLY/HOMEWORK

Homework Assignment

_____ Basic: EP p. 804 Exs. 36–50 even; pp. 123–124 Exs. 8–16, 20–24, 28–31, 37–45

_____ Average: pp. 123–124 Exs. 14–22, 25–29, 32–35, 37–46

_____ Advanced: pp. 123–124 Exs. 14–24, 27, 30–38*, 43–46

Reteaching the Lesson

_____ Practice: CRB pages 8–10 (Level A, Level B, Level C); Practice Workbook

_____ Study Guide: CRB pages 11–12; Spanish Study Guide

Extending the Lesson

_____ Real-World Problem Solving: CRB page 13

_____ Challenge: SE page 124; CRB page 14

ASSESSMENT OPTIONS

_____ Daily Quiz (3.1): TE page 124 or Transparencies

_____ Standardized Test Practice: SE page 124

Notes

____

____

LESSON **3.1** Teacher's Name ______ Class ______ Room ______ Date ______

Lesson Plan for Block Scheduling

Half-block lesson (See *Pacing and Assignment Guide*, TE page 116A)

For use with pages 119–124

GOAL **Solve two-step equations.**

State/Local Objectives ______

✓ **Check the items you wish to use for this lesson.**

Chapter Pacing Guide	
Day	**Lesson**
1	**3.1;** 3.2
2	3.3
3	3.4
4	3.5
5	3.6
6	Ch. 3 Review and Projects

STARTING OPTIONS

_____ Warm-Up: Transparencies

TEACHING OPTIONS

_____ Notetaking Guide
_____ Concept Activity: SE page 119
_____ Activity Support Master: CRB page 7
_____ Examples: 1–4, SE pages 120–122
_____ Extra Examples: TE pages 121–122
_____ Checkpoint Exercises: 1–12, SE pages 120–121
_____ Concept Check: TE page 122
_____ Guided Practice Exercises: 1–7, SE page 122

APPLY/HOMEWORK

Homework Assignment

_____ Block Schedule: pp. 123–124 Exs. 14–22, 25–29, 32–35, 37–46 (with 3.2)

Reteaching the Lesson

_____ Practice: CRB pages 8–10 (Level A, Level B, Level C); Practice Workbook
_____ Study Guide: CRB pages 11–12; Spanish Study Guide

Extending the Lesson

_____ Real-World Problem Solving: CRB page 13
_____ Challenge: SE page 124; CRB page 14

ASSESSMENT OPTIONS

_____ Daily Quiz (3.1): TE page 124 or Transparencies
_____ Standardized Test Practice: SE page 124

Notes

Name ____________________ Date ________

LESSON 3.1

Activity Support Master

For use with page 119

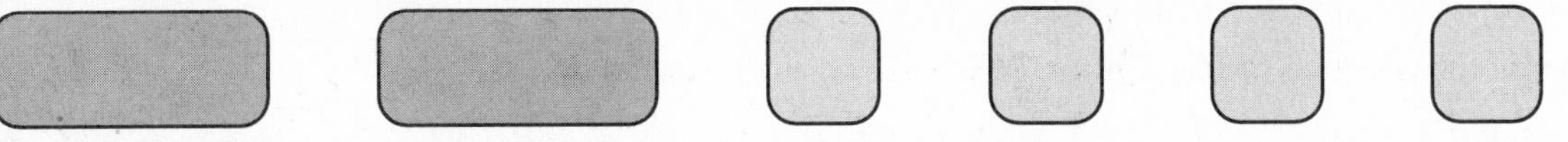
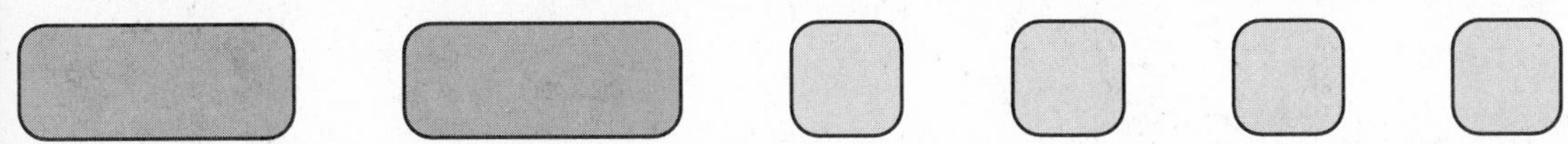
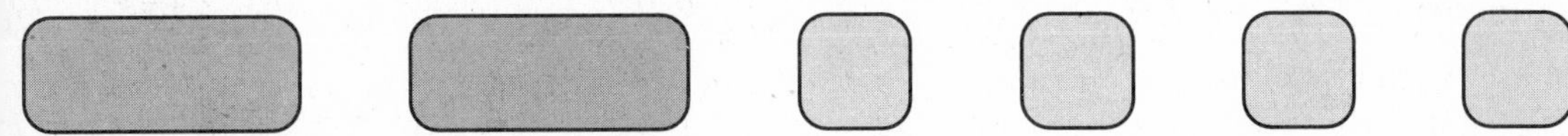

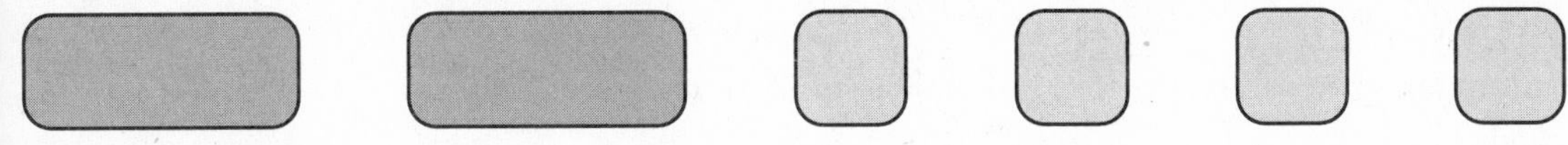
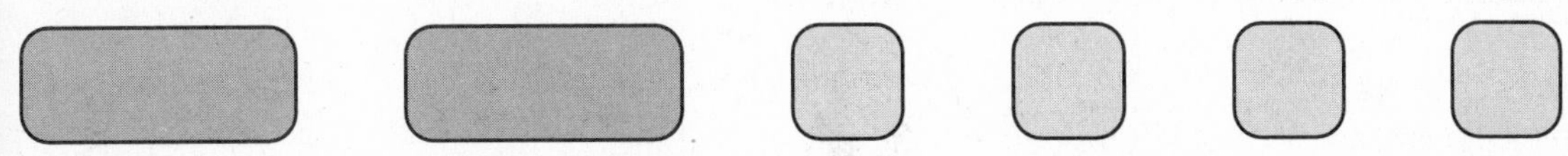
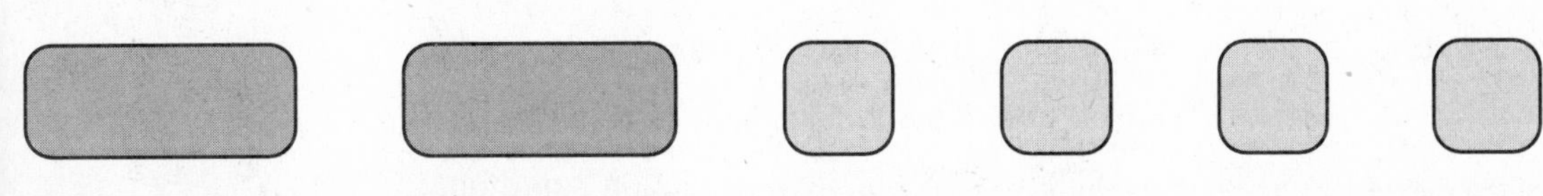

LESSON **3.1** Name ______________________ Date ________

Practice A

For use with pages 119–124

Lesson 3.1

Tell whether the given value of the variable is a solution of the equation.

1. $3x - 1 = 11; x = 4$ **2.** $1 = 2x + 7; x = -4$ **3.** $12 - x = 15; x = -3$

4. $-17 = 4x + 9; x = -2$ **5.** $-\frac{x}{5} + 7 = 5; x = 10$ **6.** $-7 = \frac{x}{6} - 10; x = 18$

Solve the equation. Check your solution.

7. $3x + 1 = 13$ **8.** $17 = 8x - 7$ **9.** $4x + 5 = 5$ **10.** $11 = 2x + 7$

11. $5x - 2 = 3$ **12.** $7x + 1 = 22$ **13.** $\frac{x}{2} - 5 = 3$ **14.** $10 = \frac{x}{4} + 7$

15. $\frac{x}{5} - 1 = 9$ **16.** $4 = \frac{x}{8} + 3$ **17.** $\frac{x}{3} + 6 = 9$ **18.** $\frac{x}{6} - 2 = 3$

19. $6 - x = 7$ **20.** $5 - 2x = 17$ **21.** $-4 = 1 - x$ **22.** $10 = 3x - 11$

23. You are buying a digital camera that costs $375. The store lets you make a down payment. You can pay the remaining cost in four equal monthly payments with no interest charged. You make a down payment of $175. Which equation can you use to find the amount of each monthly payment?

A. $375 = 175 + 4p$ **B.** $375 = 4p - 175$ **C.** $375 + 4p = 175$

24. Use the information from Exercise 23 to find the amount of each monthly payment.

25. For one day, a barber has 28 customers and receives $64 in tips. The barber charges a flat rate for haircuts and makes a total of $456 including tips. Which equation can you use to find how much the barber charges for a haircut?

A. $28x - 456 = 64$ **B.** $28x - 64 = 456$ **C.** $28x + 64 = 456$

26. Use the information from Exercise 25 to find how much the barber charges for a haircut.

27. You are building an entertainment center. The middle section of the entertainment center is 30 inches wide for your television. You also want 2 bookshelves (4 total) on each side of the middle section. The entire entertainment center is 90 inches wide. How wide can each of the bookshelves be?

a. Draw a diagram of the entertainment center. Label your diagram.

b. Write a verbal model to find the width of each bookshelf.

c. Let w represent the width of each bookshelf. Write an equation based on your verbal model.

d. Solve your equation to find the width of each bookshelf.

Name ______________________ Date __________

LESSON
3.1 Practice B

For use with pages 119–124

Tell whether the given value of the variable is a solution of the equation.

1. $6x - 7 = 17; x = 4$

2. $1 = 4x + 9; x = -2$

3. $8 - 3x = 5; x = -1$

4. $-15 = -3x + 15; x = 5$

5. $\frac{x}{5} - 6 = -2; x = 20$

6. $-6 = \frac{x}{2} - 7; x = -2$

Solve the equation. Check your solution.

7. $7x + 12 = 26$

8. $2x + 9 = -5$

9. $-4 = 9x + 23$

10. $-10 = 6x - 16$

11. $25 - 3x = -8$

12. $4x - 15 = 25$

13. $70 = 19 - 3x$

14. $-2x - 47 = -11$

15. $-14 = -22 - \frac{x}{3}$

16. $\frac{x}{12} + 13 = 18$

17. $-10 = 8 - \frac{x}{7}$

18. $3 = \frac{x}{25} + 6$

19. $250 = 124 - 3x$

20. $-\frac{x}{9} - 12 = -23$

21. $56 - \frac{x}{15} = 47$

Write the verbal sentence as an equation. Then solve the equation.

22. Fourteen minus the product of 3 and a number is 26.

23. Negative seven minus the product of 5 and number is 28.

24. Eleven minus the quotient of a number and 8 is 15.

25. Negative sixteen plus the quotient of a number and 2 is 35.

26. Thirty-nine minus a number is -19.

27. Fifteen people volunteer for a park cleanup. The number of volunteers increases by 7 people each month for several months. After how many months will there be 50 volunteers?

28. You have a $100 gift card to spend at a store. You buy a portable compact disc player for $45. Compact discs are on sale for $11 each. How many compact discs can you buy with the money remaining on the gift card?

29. A group of 4 friends are playing golf. The total cost of the round of golf is $108. Each person in the group has the same coupon. The total cost of the round with the coupons is $76. How much is the coupon worth?

30. A school makes $715 from ticket sales for a school play. From the ticket sales, $448 is from adult tickets. Student tickets are $3 each. How many students attended the play?

31. You are rock climbing and descending a cliff at a rate of about 9 feet per minute. The cliff is about 360 feet high.

a. How long until you are at a height of 234 feet?

b. How long until you are halfway down the cliff?

LESSON **3.1**

Name ______________________ Date ________

Practice C

For use with pages 119–124

Lesson 3.1

Solve the equation. Check your solution.

1. $\frac{2x}{3} - 8 = -4$

2. $-5 = \frac{5x}{8} + 25$

3. $4.3x + 8.2 = -17.6$

4. $13.8 = 2.7x - 7.8$

5. $-12.5 + 1.9x = -35.3$

6. $-6.2x - 10 = -28.6$

7. $\frac{x}{4.3} + 7.2 = 11.2$

8. $5.6 = \frac{x}{2.5} - 5.4$

9. $-4 - \frac{x}{0.6} = -24$

10. $14.6 - \frac{x}{2.8} = 7.1$

11. $13.9 = \frac{x}{6} + 12.5$

12. $\frac{x + 8}{5} = -3$

13. $-15 = \frac{x - 12}{2}$

14. $4.25x - 2.91 = 48.09$

15. $25.89 = -3.74x + 7.19$

Write the verbal sentence as an equation. Then solve the equation.

16. The product of 2.5 and a number minus 4.7 is 10.3.

17. The product of 8.3 and a number plus 5.6 is -27.6.

18. The quotient of a number and 1.8 minus 9.4 is 5.6.

19. Seven and one tenth plus the quotient of a number and 0.9 is 16.1.

20. You are saving money to buy a DVD recorder. The DVD recorder costs $380. You have already saved $170. You can save an additional $30 each month.

a. Write a variable expression to represent the total amount of money you have saved after *m* months. Evaluate your expression for the first 6 months. Record your results in a table.

b. Use the data in part (a) to make a scatter plot. Put months on the horizontal axis and savings on the vertical axis. Use the graph to find the number of months it will take to save enough money for the recorder.

c. Write and solve an equation to find the number of months it will take you to save enough money for the recorder.

21. All three sides of the triangle shown are equal in length.

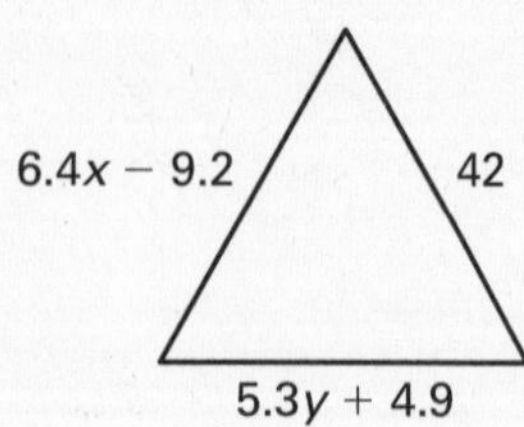

a. Write two equations you can use to find the values of the variables.

b. Find the value of each variable.

LESSON **3.1**

Name ______________________ Date __________

Study Guide

For use with pages 119–124

GOAL **Solve two-step equations.**

EXAMPLE 1 Using Subtraction and Division to Solve

Solve $14x + 12 = 54$. Check your solution.

$14x + 12 = 54$	Write original equation.
$14x + 12 - 12 = 54 - 12$	Subtract 12 from each side.
$14x = 42$	Simplify.
$\frac{14x}{14} = \frac{42}{14}$	Divide each side by 14.
$x = 3$	Simplify.

Answer: The solution is 3.

✓ Check $14x + 12 = 54$	Write original equation.
$14(3) + 12 \stackrel{?}{=} 54$	Substitute 3 for x.
$54 = 54$ ✓	Solution checks.

Exercises for Example 1

Solve the equation. Check your solution.

1. $12x + 15 = 75$ **2.** $13n + 3 = 81$ **3.** $5z + 20 = 95$ **4.** $5x + 11 = 111$

EXAMPLE 2 Using Addition and Multiplication to Solve

Solve $\frac{x}{6} - 8 = 6$.

$\frac{x}{6} - 8 = 6$	Write original equation.
$\frac{x}{6} - 8 + 8 = 6 + 8$	Add 8 to each side.
$\frac{x}{6} = 14$	Simplify.
$6\left(\frac{x}{6}\right) = 6(14)$	Multiply each side by 6.
$x = 84$	Simplify.

Answer: The solution is 84.

Lesson 3.1

Name ______________________ Date ________

Study Guide

For use with pages 119–124

Exercises for Example 2

Solve the equation. Check your solution.

5. $\frac{y}{10} - 13 = -27$

6. $\frac{m}{7} - 8 = 1$

7. $\frac{g}{12} - 5 = 3$

8. $\frac{r}{2} - 5 = 40$

EXAMPLE 3 Solving an Equation with Negative Coefficients

Solve $10 - \frac{n}{5} = -3$. Check your solution.

$10 - \frac{n}{5} = -3$	Write original equation.
$10 - \frac{n}{5} - 10 = -3 - 10$	Subtract 10 from each side.
$-\frac{n}{5} = -13$	Simplify.
$\frac{n}{-5} = -13$	Rewrite $-\frac{n}{5}$ as $\frac{n}{-5}$.
$-5\left(\frac{n}{-5}\right) = -5(-13)$	Multiply each side by -5.
$n = 65$	Simplify.

Answer: The solution is 65.

✓ **Check**	$10 - \frac{n}{5} = -3$	Write original equation.
	$10 - \frac{65}{5} \stackrel{?}{=} -3$	Substitute 65 for n.
	$-3 = -3$ ✓	Solution checks.

Exercises for Example 3

Solve the equation. Check your solution.

9. $10 - 15y = 55$

10. $11 - 9n = -16$

11. $21 - \frac{x}{2} = 7$

12. $2 - \frac{p}{11} = 11$

LESSON 3.1

Name ____________________ Date __________

Real-World Problem Solving

For use with pages 119–124

Buying a Car

When you negotiate with an automobile dealer to buy a car, the purchase price you agree to is not all that you have to pay to drive that car off the lot. There are extra costs such as sales tax, title fee, registration fee, license plate fee, and other miscellaneous charges that are added to the purchase price.

In Exercises 1–6, use the following information.

You take a job doing paperwork for a car dealer. One of your duties is to find a customer's total cost for an automobile when it is offered for a specified purchase price. To do this you need to find the extra costs to the customer. The extra costs are the sales tax (6 percent of the price of the car), the title fee ($22.50), the registration fee ($36.00), the license fee ($12.00), and other miscellaneous fees ($17.50). You decide to write a formula for the extra costs.

1. Find the total amount of the title fee, registration fee, license fee, and miscellaneous fees for a car.

2. Write an expression for the sales tax of a car. Use 0.06 as the tax rate and let p represent the purchase price of the car.

3. Write a verbal model that relates the extra costs to the sales tax and the total amount of the fees.

4. Write a formula based on the verbal model. Let C represent the extra costs.

5. Use the formula to find a customer's extra costs for a car with a purchase price of $15,780.

6. You have misplaced some paperwork on a price offer made to a customer. You find a document showing that the extra costs for the purchase price would be $1243. Find the purchase price offered to the customer.

LESSON **3.1**

Name ______________________________ Date __________

Challenge Practice

For use with pages 119–124

Solve the equation. Check your solution.

1. $-8.2a + 6.7 = -5.6$

2. $-12.9 + 7.4x = 35.2$

3. $3.5 + \frac{m}{7.5} = 7.7$

4. $3.9 - \frac{c}{2.2} = -0.2$

5. A CD club is selling CDs for $7.50 each. The charge for shipping the CDs to your home is $5. How many CDs can you buy for $50?

6. You are at a book sale where all paperback books are $3 each. You have $25, but you need $2 to ride the bus home. How many paperback books can you buy at the sale?

Write the verbal sentence as an equation. Then solve the equation.

7. Eighteen minus the quotient of a number and 4 is 54.

8. The quotient of a number and 8 subtracted from 25 is 20.

9. Solve $7 = \frac{x - 4}{5}$. Explain how you solved the equation and how you know your solution is correct.

Lesson 3.1

LESSON **3.2** Teacher's Name ______________________ Class ______ Room ______ Date ______

Lesson Plan

1-day lesson (See *Pacing and Assignment Guide*, TE page 116A)

For use with pages 125–129

GOAL **Solve equations using the distributive property.**

State/Local Objectives ____

____

✓ Check the items you wish to use for this lesson.

STARTING OPTIONS

_____ Homework Check (3.1): TE page 123; Answer Transparencies

_____ Homework Quiz (3.1): TE page 124; Transparencies

_____ Warm-Up: Transparencies

TEACHING OPTIONS

_____ Notetaking Guide

_____ Activity Master: CRB page 17

_____ Examples: 1–3, SE pages 125–126

_____ Extra Examples: TE page 126

_____ Checkpoint Exercises: 1–3, SE page 126

_____ Concept Check: TE page 126

_____ Guided Practice Exercises: 1–9, SE page 127

APPLY/HOMEWORK

Homework Assignment

_____ Basic: EP p. 804 Exs. 20–23; pp. 127–129 Exs. 11–20, 22–27, 33–35, 40–55

_____ Average: pp. 127–129 Exs. 10–16, 20–25, 30–38, 42–45, 50–56

_____ Advanced: pp. 127–129 Exs. 10–13, 17–29, 32–39*, 44–47, 52–56

Reteaching the Lesson

_____ Practice: CRB pages 18–20 (Level A, Level B, Level C); Practice Workbook

_____ Study Guide: CRB pages 21–22; Spanish Study Guide

Extending the Lesson

_____ Challenge: SE page 129; CRB page 23

ASSESSMENT OPTIONS

_____ Daily Quiz (3.2): TE page 129 or Transparencies

_____ Standardized Test Practice: SE page 129

Notes

____

____

LESSON **3.2**

Teacher's Name ______________________ Class ______ Room ______ Date ______

Lesson Plan for Block Scheduling

Half-block lesson (See *Pacing and Assignment Guide*, TE page 116A)

For use with pages 125–129

GOAL **Solve equations using the distributive property.**

State/Local Objectives ____

____

✓ **Check the items you wish to use for this lesson.**

Chapter Pacing Guide	
Day	**Lesson**
1	3.1; **3.2**
2	3.3
3	3.4
4	3.5
5	3.6
6	Ch. 3 Review and Projects

STARTING OPTIONS

_____ Homework Check (3.1): TE page 123; Answer Transparencies

_____ Homework Quiz (3.1): TE page 124; Transparencies

_____ Warm-Up: Transparencies

TEACHING OPTIONS

_____ Notetaking Guide

_____ Activity Master: CRB page 17

_____ Examples: 1–3, SE pages 125–126

_____ Extra Examples: TE page 126

_____ Checkpoint Exercises: 1–3, SE page 126

_____ Concept Check: TE page 126

_____ Guided Practice Exercises: 1–9, SE page 127

Lesson 3.2

APPLY/HOMEWORK

Homework Assignment

_____ Block Schedule: pp. 127–129 Exs. 10–16, 20–25, 30–38, 42–45, 50–56 (with 3.1)

Reteaching the Lesson

_____ Practice: CRB pages 18–20 (Level A, Level B, Level C); Practice Workbook

_____ Study Guide: CRB pages 21–22; Spanish Study Guide

Extending the Lesson

_____ Challenge: SE page 129; CRB page 23

ASSESSMENT OPTIONS

_____ Daily Quiz (3.2): TE page 129 or Transparencies

_____ Standardized Test Practice: SE page 129

Notes

____

____

____

____

LESSON **3.2** Name ______________________ Date __________

Activity Master

For use before Lesson 3.2

Goal
Solve equations using algebra tiles.

Materials
- algebra tiles
- pencil and paper

Modeling Equations Having Like Terms

In this activity, you will model and solve equations using algebra tiles.

INVESTIGATE **Use algebra tiles to solve $2x + 3 + x = 12$.**

1 Model $2x + 3 + x = 12$ using algebra tiles.

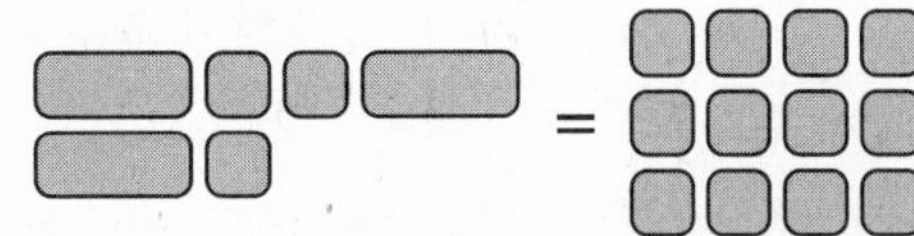

2 Group like tiles.

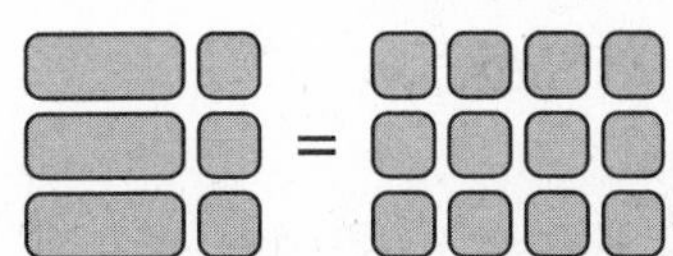

3 Remove three 1-tiles from each side.

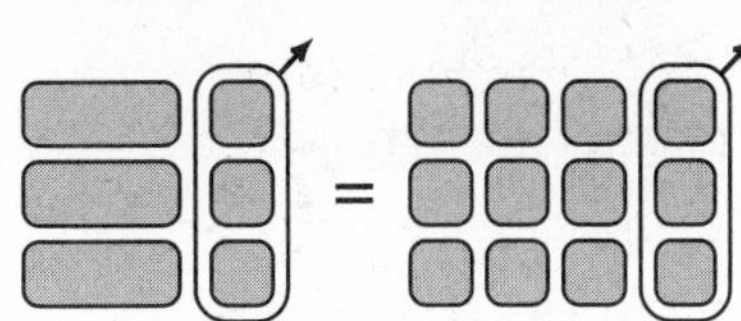

4 Divide the remaining tiles into three equal groups. Each x-tile is equal to three 1-tiles. So, the solution is 3.

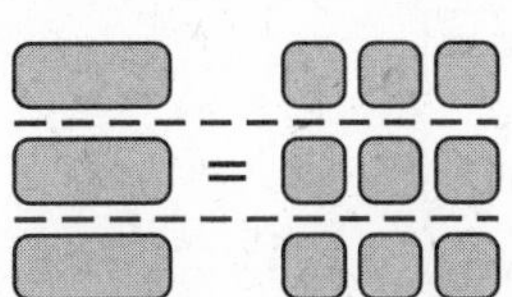

DRAW CONCLUSIONS

Use algebra tiles to solve the equation.

1. $x + 2 + 4x = 7$ **2.** $2x + 5 + 2x = 13$ **3.** $3x + 3 + x = 15$

4. $x + 3 + 2x = 15$ **5.** $26 = x + 6 + 4x$ **6.** $4 + x + 2 + 2x = 12$

7. Describe the steps you would take to solve $x + 5 + 2x = 11$ without using algebra tiles.

LESSON **3.2** Name ______________________________ Date __________

Practice A

For use with pages 125–129

Tell whether the given value of the variable is a solution of the equation.

1. $5 + x + 3 = 14; x = 7$

2. $4x + 9x + 1 = 27; x = 2$

3. $4(x + 7) = 16; x = -3$

4. $18 - 2(x - 3) = 26; x = 1$

Solve the equation. Check your solution.

5. $3x + 8 + x = 28$

6. $9x + 7 - 2x = 14$

7. $17 + 3x - 11 = 0$

8. $12x - 1 - 10x = 23$

9. $3(6 - x) = 27$

10. $-2(x + 7) = 16$

11. $-40 = 4(x - 10)$

12. $20 = -5(x + 7)$

13. $-6(2x + 3) = 42$

14. $2(11 - 4x) = 38$

15. $14x - 6 - 11x = 21$

16. $8 + 5x - 6 = 37$

Find the value of *x* for the given triangle, rectangle, or square.

17. Perimeter = 30 units

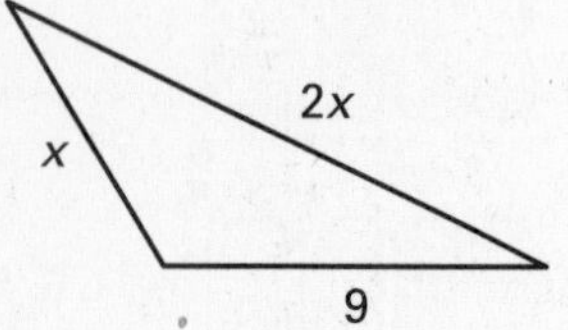

18. Perimeter = 24 units

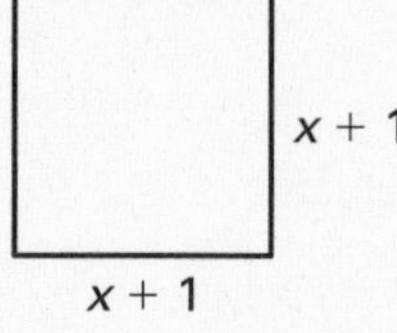

19. Perimeter = 20 units

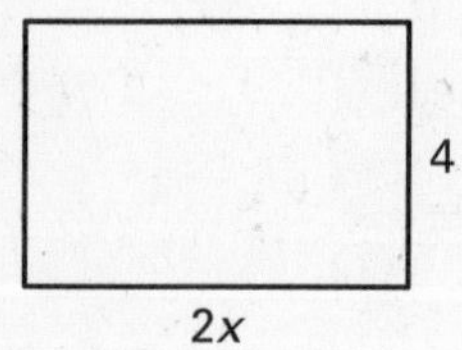

20. Perimeter = 50 units

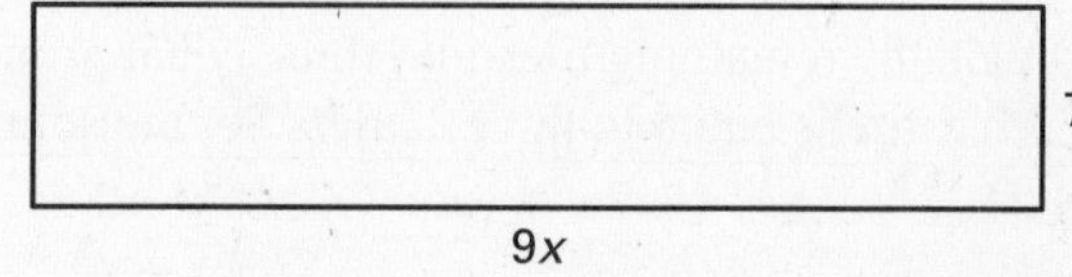

21. The perimeter of a rectangular picture frame is 30 inches. The length of the frame is three less than twice the width. What are the dimensions of the frame?

22. You spend \$91 shopping for new clothes. You spend \$24 for a pair jeans and \$35 for a pair of shoes. You also buy 4 shirts that each cost *d* dollars. How much is each shirt?

23. You spend \$55 shopping for birthday gifts. You buy one of your friends a gift certificate for \$15, and your other friend a pair of shorts for \$16. You also buy your older brother 2 compact discs that each cost *d* dollars. How much is each compact disc?

LESSON **3.2**

Name ______________________________ Date __________

Practice B

For use with pages 125–129

Solve the equation. Check your solution.

1. $10 + 3(x + 2) = 31$

2. $-2(x - 6) + 7 = 35$

3. $-20 - (4x - 1) = -15$

4. $12(x + 3) - 3x = 117$

5. $-25 + 4(2x + 5) = -61$

6. $187 = 19 + 7(13 - x)$

7. $20 = 14 + 3(x + 8)$

8. $-5(2x - 7) + 24 = 89$

9. $-14 = 6x - 8(x + 3)$

10. $-7x - (10 - x) = -58$

11. $48 = 15 + 6(4 + x) - 3x$

12. $23 - 7(x + 3) + 5x = 10$

Find the value of *x* for the given triangle, rectangle, or square.

13. Perimeter = 29 units

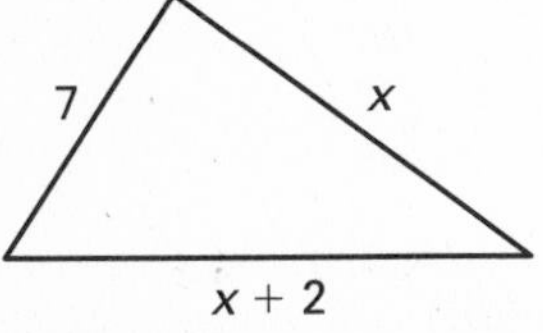

14. Perimeter = 28 units

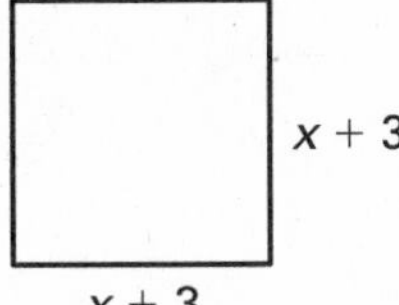

15. Perimeter = 52 units

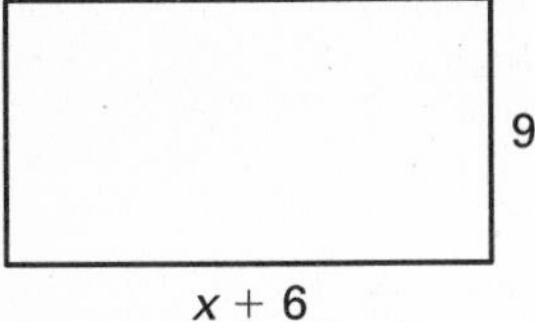

16. Perimeter = 38 units

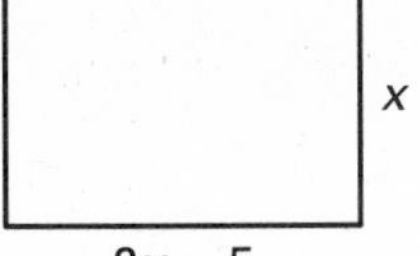

17. The length of a rectangle is 3 meters more than twice its width. The perimeter of the rectangle is 48 meters. Let w represent the width.

a. Sketch a diagram of the rectangle.

b. Write an equation for the perimeter of the rectangle.

c. Find the length and width of the rectangle.

18. A class of 42 students and 2 teachers plan a trip to an observatory. The class has raised \$485 for the trip. Admission is \$5 per person and bus rental is \$230. With the remaining money, the class can invite guests to fill the remaining seats on the bus. Write and solve an equation to find the number of guests g the class can invite.

19. A plumber charges \$30 per hour and \$42 for each hour of overtime. For a job, the plumber works 3 regular hours, h overtime hours, and charges \$195 for new parts. The total amount of the bill for the job is \$390. Write and solve an equation to find the number of overtime hours the plumber worked.

Lesson 3.2

LESSON **3.2** Name ______________________ Date __________

Practice C

For use with pages 125–129

Solve the equation. Check your solution.

1. $-7x - (13 - x) = 11$

2. $119 = 35 + 4(6 - x)$

3. $-26 - (7x + 8) = -55$

4. $-10x - 7(5 - 2x) = -71$

5. $18x + 6(4 - x) = 72$

6. $-2 = 11 + 3(3x - 5) - 7x$

7. $6 + 0.5(x - 2) = 9$

8. $23 = 6 + 0.25(8 - 12x)$

9. $6x + 3.5(x - 14) = 8$

10. $32 = 16 + 0.4(20 - 10x)$

11. $19 - 0.25(x + 8) - x = 2$

12. $2.4(x + 15) + 7.3 = 67.3$

Find the value of *x* for the given triangle, rectangle, or square.

13. Perimeter = 19.8 units

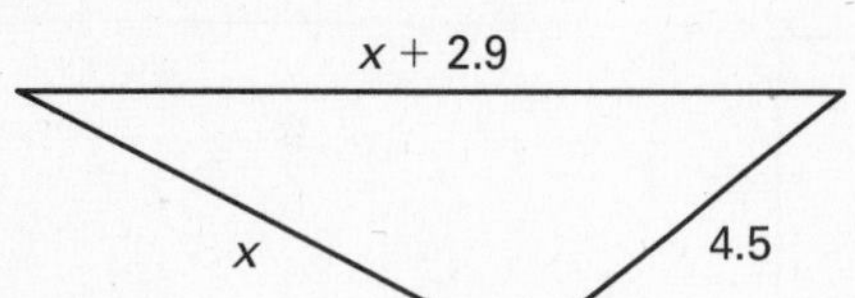

14. Perimeter = 66.8 units

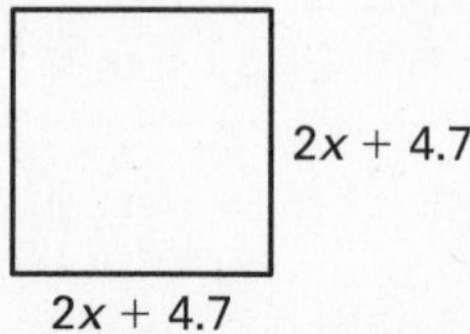

15. Perimeter = 108 units

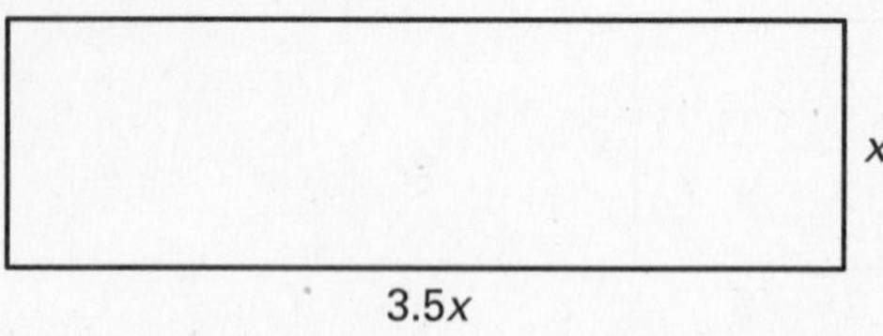

16. Perimeter = 38.5 units

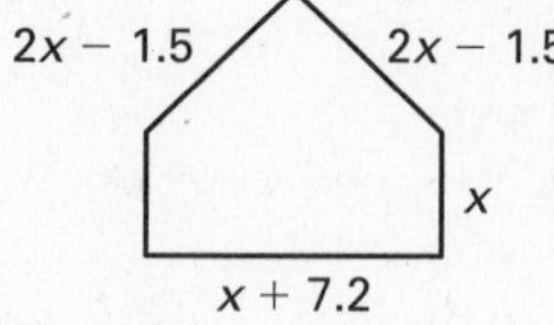

Lesson 3.2

17. Kevin, Mika, and Cheryl have a total of 45 people sponsoring them for a fund-raiser. Kevin has p people sponsoring him, Mika has 1 more than Kevin, and Cheryl has 1 more than Mika. Write and solve an equation to find the number of people sponsoring each person.

18. You spend \$100.39 shopping for new clothes. You spend \$28.90 for a pair of jeans and \$49.99 for a pair of sneakers. You also buy 2 shirts that each cost d dollars. How much is each shirt?

19. Consecutive integers are 1, 2, 3, 4, and so on. Consecutive even integers are 2, 4, 6, 8, and so on. The sum of an even integer and twice the next consecutive even integer is 178. Write and solve an equation to find the two integers. Let $2x$ represent the first even integer.

20. Consecutive odd integers are 1, 3, 5, 7, and so on. The sum of two consecutive odd integers is 104. Write and solve an equation to find the two integers. Let $2x - 1$ represent the first odd integer.

Name ______________________ Date __________

LESSON 3.2

Study Guide

For use with pages 125–129

GOAL Solve equations using the distributive property.

EXAMPLE 1 Writing and Solving an Equation

You are giving a birthday party. For the party, you want to buy a personalized birthday banner that costs \$17, foil balloons that cost \$1.90 each, and packs of streamers that cost \$1.10 each. You have a total budget of \$35. If you buy equal numbers of foil balloons and packs of streamers, how many can you afford to buy?

Solution

Let n represent the number of foil balloons and the number of packs of streamers. Then $1.90n$ represents the cost of n balloons, and $1.10n$ represents the cost of n packs of streamers. Write a verbal model.

Cost of foil balloons	+	Cost of packs of streamers	+	Cost of banner	=	Total budget

$1.90n + 1.10n + 17 = 35$	Substitute.
$3n + 17 = 35$	Combine like terms.
$3n + 17 - 17 = 35 - 17$	Subtract 17 from each side.
$3n = 18$	Simplify.
$\frac{3n}{3} = \frac{18}{3}$	Divide each side by 3.
$n = 6$	Simplify.

Answer: You can afford to buy 6 foil balloons and 6 packs of streamers.

Exercise for Example 1

1. You have \$15.50 to spend on party food. Gum drops are \$3 per pound, yogurt-covered peanuts are \$2.50 per pound, and cashews are \$4.50 per pound. If you buy 0.5 pound of gum drops and an equal weight of yogurt-covered peanuts and cashews, how much can you afford to buy?

EXAMPLE 2 Solving Equations Using the Distributive Property

Solve $-8(5 - 7c) = 184$.

$-8(5 - 7c) = 184$	Write original equation.
$-40 + 56c = 184$	Distributive property
$-40 + 56c + 40 = 184 + 40$	Add 40 to each side.
$56c = 224$	Simplify.
$\frac{56c}{56} = \frac{224}{56}$	Divide each side by 56.
$c = 4$	Simplify.

Name ______________________________ Date __________

LESSON 3.2 Continued

Study Guide

For use with pages 125–129

Exercises for Example 2

Solve the equation. Check your solution.

2. $-2(7 - 11v) = 96$ **3.** $12(4 - 3g) = -132$ **4.** $-5(c - 1) = -50$

EXAMPLE 3 Combining Like Terms After Distributing

Solve $13x - (x + 7) = 29$.

$13x - (x + 7) = 29$	Write original equation.
$13x - x - 7 = 29$	Distributive property
$12x - 7 = 29$	Combine like terms.
$12x - 7 + 7 = 29 + 7$	Add 7 to each side.
$12x = 36$	Simplify.
$\frac{12x}{12} = \frac{36}{12}$	Divide each side by 12.
$x = 3$	Simplify.

Exercises for Example 3

Solve the equation. Check your solution.

5. $10a + 5(2a + 1) = 65$ **6.** $18b + 8(3b + 7) = 98$ **7.** $11z - 3(z - 9) = 123$

LESSON 3.2

Name ____________________ Date __________

Challenge Practice

For use with pages 125-129

Solve the equation. Check your solution.

1. $61 = 16 + 5(t - 3)$
2. $-8(3x - 4) + 1 = 15$
3. $8c = 14 - 0.5(10c + 2)$
4. $4z = 28 - 0.25(4 - 16z)$
5. Your brother bought some fish and an aquarium for $250. The aquarium was $175 and the fish were $6.25 each. How many fish did your brother buy?
6. The sum of three numbers is 39. Let x be the first number. The second number is 5 more than the first number and the third number is two times the second number. What are the three numbers?
7. The difference of two numbers is -14. Let m be the first number. The second number is 4 more than 3 times the first number. What are the two numbers?

Find the value of *x*.

8. The figure below is composed of a rectangle and two congruent triangles. It has an area of 90 square units.

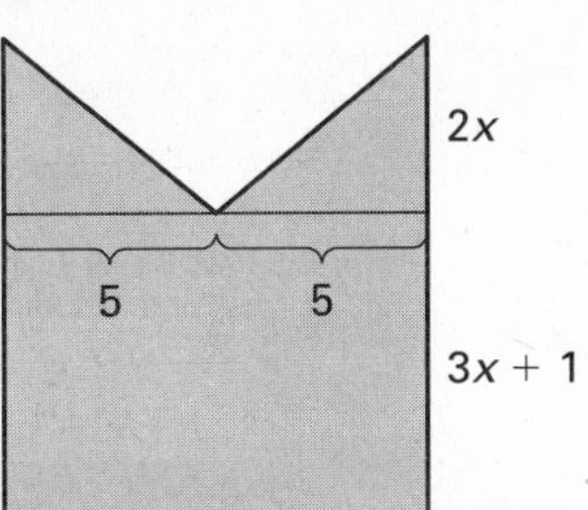

9. The figure below is composed of a rectangle and two pairs of congruent triangles. It has an area of 19 square units.

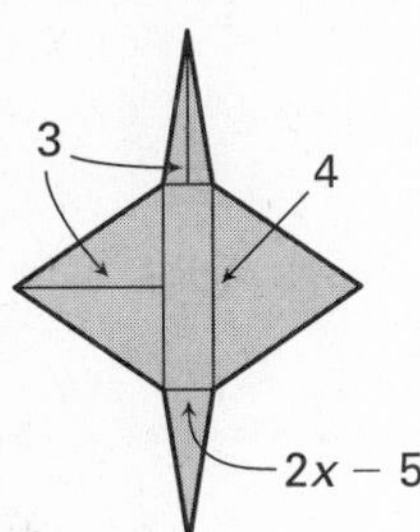

Lesson 3.2

LESSON **3.3**

Teacher's Name ____________________ Class ______ Room ______ Date ______

Lesson Plan

2-day lesson (See *Pacing and Assignment Guide*, TE page 116A)

For use with pages 130–136

GOAL **Solve equations with variables on both sides.**

State/Local Objectives ____

____

✓ Check the items you wish to use for this lesson.

STARTING OPTIONS

____ Homework Check (3.2): TE page 127; Answer Transparencies

____ Homework Quiz (3.2): TE page 129; Transparencies

____ Warm-Up: Transparencies

TEACHING OPTIONS

____ Notetaking Guide

____ Concept Activity: SE page 130; Activity Support Master: CRB page 26

____ Examples: Day 1: 1–2, SE pages 131–132; Day 2: 3–5, SE pages 132–133

____ Extra Examples: TE pages 132–133

____ Checkpoint Exercises: Day 1: 1–3, SE page 131; Day 2: none

____ Concept Check: TE page 133

____ Guided Practice Exercises: Day 1:1, 3–10, SE page 133; Day 2: 2, SE page 133

____ Technology Activity: SE page 136; Technology Keystrokes: CRB page 28

APPLY/HOMEWORK

Homework Assignment

____ Basic: Day 1: SRH p. 785 Exs. 1, 5; pp. 134–135 Exs. 11–14, 23–27, 34–37, 44–47
Day 2: pp. 134–135 Exs. 15–22, 28–30, 48–51

____ Average: Day 1: pp. 134–135 Exs. 11–14, 25–27, 31–35, 44–47
Day 2: pp. 134–135 Exs. 17–22, 28–30, 38–40, 48–51

____ Advanced: Day 1: pp. 134–135 Exs. 11–14, 25–27, 31–33, 39–45*
Day 2: pp. 134–135 Exs. 17–22, 28–30, 36–38, 46–51

Reteaching the Lesson

____ Practice: CRB pages 29–31 (Level A, Level B, Level C); Practice Workbook

____ Study Guide: CRB pages 32–33; Spanish Study Guide

Extending the Lesson

____ Challenge: SE page 135; CRB page 34

ASSESSMENT OPTIONS

____ Daily Quiz (3.3): TE page 135 or Transparencies

____ Standardized Test Practice: SE page 135

____ Quiz (3.1–3.3): SE page 137; Assessment Book page 29

Notes ____

LESSON **3.3**

Teacher's Name ______________________ Class ______ Room ______ Date ______

Lesson Plan for Block Scheduling

1-block lesson (See *Pacing and Assignment Guide*, TE page 116A)

For use with pages 130–136

GOAL **Solve equations with variables on both sides.**

State/Local Objectives ____

____

✓ Check the items you wish to use for this lesson.

Chapter Pacing Guide	
Day	**Lesson**
1	3.1; 3.2
2	**3.3**
3	3.4
4	3.5
5	3.6
6	Ch. 3 Review and Projects

STARTING OPTIONS

____ Homework Check (3.2): TE page 127; Answer Transparencies

____ Homework Quiz (3.2): TE page 129; Transparencies

____ Warm-Up: Transparencies

TEACHING OPTIONS

____ Notetaking Guide

____ Concept Activity: SE page 130;
Activity Support Master: CRB page 26

____ Examples: 1–5, SE pages 131–133

____ Extra Examples: TE pages 132–133

____ Checkpoint Exercises: 1–3, SE page 131

____ Concept Check: TE page 133

____ Guided Practice Exercises: 1–10, SE page 133

____ Technology Activity: SE page 136; Technology Keystrokes: CRB page 28

APPLY/HOMEWORK

Homework Assignment

____ Block Schedule: pp. 134–135 Exs. 11–14, 17–22, 25–35, 38–40, 44–51

Reteaching the Lesson

____ Practice: CRB pages 29–31 (Level A, Level B, Level C); Practice Workbook

____ Study Guide: CRB pages 32–33; Spanish Study Guide

Extending the Lesson

____ Challenge: SE page 135; CRB page 34

ASSESSMENT OPTIONS

____ Daily Quiz (3.3): TE page 135 or Transparencies

____ Standardized Test Practice: SE page 135

____ Quiz (3.1–3.3): SE page 137; Assessment Book page 29

Notes

____

____

LESSON 3.3

Name ______________________________ Date __________

Activity Support Master

For use with page 130

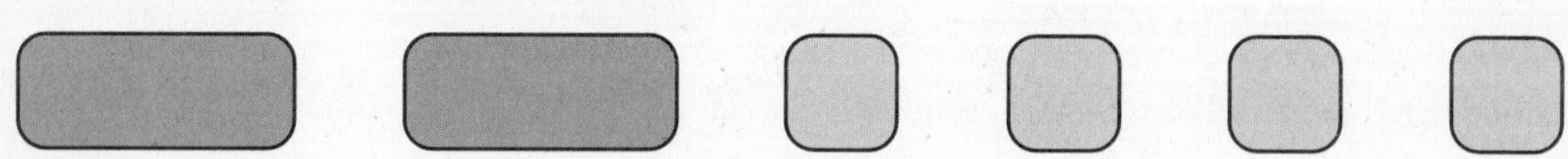

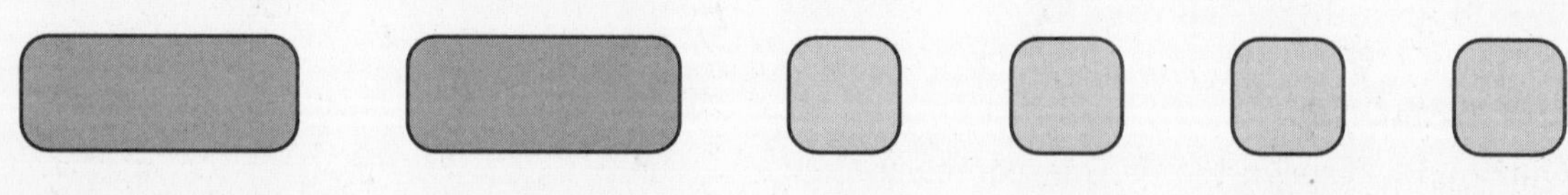

Lesson 3.3

LESSON 3.3

Name ______________________________ Date __________

Technology Keystrokes

For use with Exercises 39–42, page 135

TI-34 II

39. [(–)] 10.3936 [–] 14 [ENTER] [÷] [(] 0.75 [–] 1.87 [)] [ENTER]

40. 10.6206 [–] 19.5 [ENTER] [÷] [(] 0.5 [+] 0.4 [)] [ENTER]

41. 11.08 [+] 9.39 [ENTER] [÷] [(] [(–)] 3.4 [+] 1.1 [)] [ENTER]

42. [(–)] 55.104 [+] 130.5 [ENTER] [÷] [(] [(–)] 9 [–] 3.2 [)] [ENTER]

TI-73 Explorer

39. [(–)] 10.3936 [–] 14 [ENTER] [÷] [(] 0.75 [–] 1.87 [)] [ENTER]

40. 10.6206 [–] 19.5 [ENTER] [÷] [(] 0.5 [+] 0.4 [)] [ENTER]

41. 11.08 [+] 9.39 [ENTER] [÷] [(] [(–)] 3.4 [+] 1.1 [)] [ENTER]

42. [(–)] 55.104 [+] 130.5 [ENTER] [÷] [(] [(–)] 9 [–] 3.2 [)] [ENTER]

LESSON 3.3

Name ______________________________ Date __________

Technology Activity Keystrokes

For use with Technology Activity 3.3, page 136

TI-73 Explorer

Enter the expressions on each side of the equal sign into the calculator.

[Y=] 5 [x] [−] 1 [ENTER] 4 [x] [+] 3

Enter the table settings.

[2nd] [TBLSET] 0 [ENTER] 1 [ENTER] [ENTER] [▼] [ENTER] [2nd] [TABLE]

Use the down arrow key, [▼], to scroll down the table.

Name ______________________________ Date __________

LESSON **3.3**

Practice A

For use with pages 130–136

Tell whether the given value of the variable is a solution of the equation.

1. $8x = 6x - 20;\ x = -10$
2. $6x - 1 = 3x + 8;\ x = -3$
3. $-3x - 13 = -7x + 15;\ x = -7$
4. $-2x + 5 = 7x - 22;\ x = 3$

Solve the equation. Check your solution.

5. $9x = 7x + 22$
6. $14x - 3 = 10x + 1$
7. $6x + 5 = 4x - 9$
8. $10 + 3x = 26 - 5x$
9. $3(4x - 1) = 12x$
10. $11 - 2x = 31 - 7x$
11. $9x - 10 = 5x + 14$
12. $16x + 21 = 30 + 13x$
13. $-8x - 1 = -5x + 23$
14. $4x + 10 = 2(2x + 5)$
15. $12x - 7 = 5x + 49$
16. $-4x + 10 = 6x - 40$

Write the verbal sentence as an equation. Then solve the equation.

17. Five minus 6 times a number is equal to -11 plus 2 times the number.
18. Four less than -7 times a number is equal to 13 minus 6 times the number.
19. Eight times a number plus 5 is equal to 5 times the number minus 13.
20. One less than 10 times a number is equal to -2 times the number plus 35.

Find the value of *x* for the given square.

21.

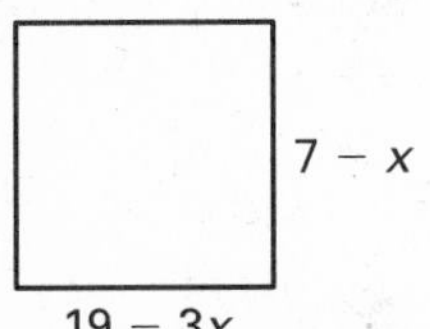

22.

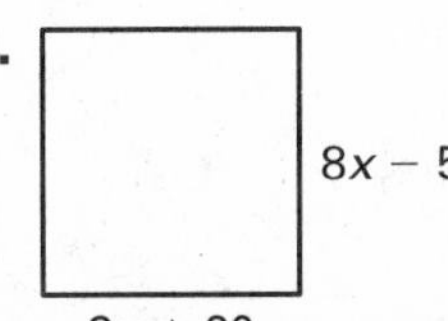

23. You and your brother are saving money to buy a camcorder. You already have \$60 saved and your brother has \$45 saved. You plan on saving an additional \$5 each week. Your brother plans on saving an additional \$8 each week. Write and solve an equation to find how many weeks it takes both of you to save the same amount. Let w represent the number of weeks.

24. The length of a football field including the end zones is 48 feet longer than four times the length of a tennis court. It is also 282 feet longer than a tennis court. Write and solve an equation to find the length (in feet) of a tennis court and a football field. Let t represent the length of a tennis court.

LESSON **3.3**

Name ______________________________ Date __________

Practice B

For use with pages 130–136

Tell whether the given value of the variable is a solution of the equation.

1. $41 - 8x = -6x - 23;\ x = -9$

2. $4x + 13 = -9 - 3(x + 9);\ x = -7$

3. $-2(3x + 7) = -3(2x + 8);\ x = -5$

4. $-9x + 7 = 25 + 2(5 - x);\ x = -4$

Solve the equation. Check your solution.

5. $12x - 28 = -63 + 7x$

6. $6x - 21 = 33 + 9x$

7. $-15x = -5(3x + 7)$

8. $16x - 19 = 113 - 6x$

9. $-19x - 34 = 56 - x$

10. $-6(4x + 3) = 6(-4x - 3)$

11. $3(-2x + 5) = 11 - 4x$

12. $14 - 9x = -8(10 + x)$

13. $-3(8x + 11) = 6(-4x - 13)$

14. $5x - 8 = 13 + 7(x - 3)$

15. $15x + 24 = 8(10 + 3x) - 2$

16. $-9x + 15 = -22 - 4(x + 12)$

Write the verbal sentence as an equation. Then solve the equation.

17. Negative thirteen times a number plus 20 is equal to -11 times the number plus 38.

18. Seventeen less than 6 times a number is equal to 47 plus 10 times the number.

19. Twenty nine less than -10 times a number is equal to -18 times the number plus 91.

20. Seventeen times a number minus 56 is equal to 10 times the number minus 63.

Find the perimeter of the triangle or rectangle. The sides of the triangle are equal in length.

21.

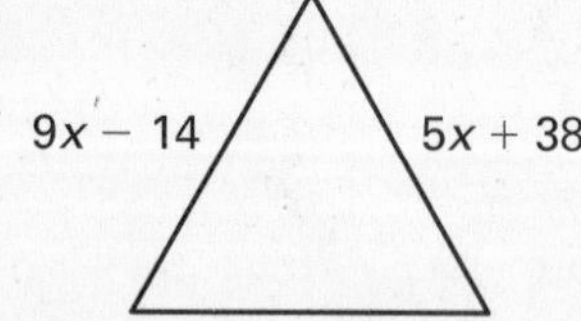

22.

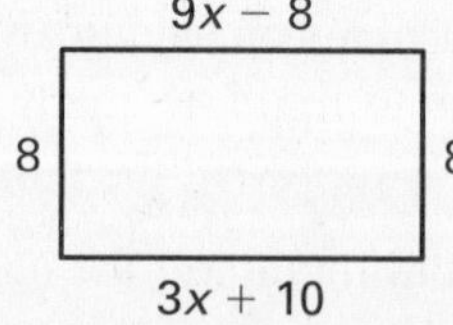

23. You are buying flowers to hand out at a school dance. Roses cost $30 for a dozen but cost more if bought individually. With the money you have, you can buy 7 dozen and 4 single roses, or 64 single roses. How much is one rose? How much money do you have?

24. The populations of two towns are changing at steady rates. One town has a population of 25,500. Its population is increasing by 2000 people each year. The other town has a population of 47,900. Its population is decreasing by 800 people each year. If the rates for each town remain the same, in how many years will the populations be the same?

LESSON

3.3 Practice C

Name ______________________________ Date __________

For use with pages 130-136

Solve the equation. Check your solution.

1. $10x - 43 = 39 + 7(x - 10)$
2. $9x - 4(x + 8) = 3(2x - 7)$
3. $-3x - 5(2x - 5) = 6(-3x - 10)$
4. $0.6x - 14 = 0.5x - 6.5$
5. $0.75x + 8 = -20 + 0.25x$
6. $-3.3x + 9.2 = -6.8 + 12.7x$
7. $5.1x - 11.4 = -0.9x - 17.4$
8. $24.4 + 2.7x = 5x - 30.8$
9. $4(x - 0.5) + 10.5 = 2(x + 11.5) - 2.5$
10. $-3(2.5x - 1) + 20.2 = -35.475 - 5x$

Write the verbal sentence as an equation. Then solve the equation.

11. Six plus 0.25 times a number is equal to 0.8 times the number minus 5.
12. Twelve minus 1.8 times a number is equal to 22.8 minus 2.7 times the number.
13. Negative twenty minus 4.6 times a number is equal to -2.3 times the number plus 16.8.
14. Seven times a number minus 10.75 is equal to 6.8 times the number plus 0.95.

Find the perimeter of the triangle, rectangle, or square. The sides of the triangle are equal in length.

15.

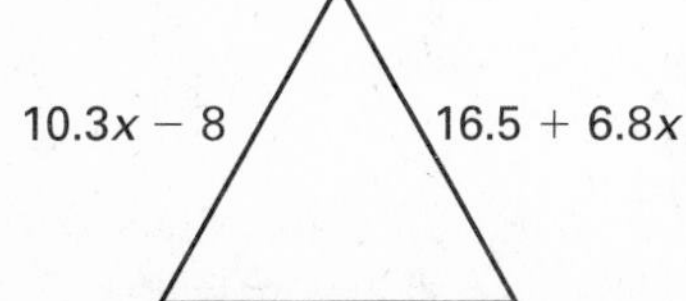

16.

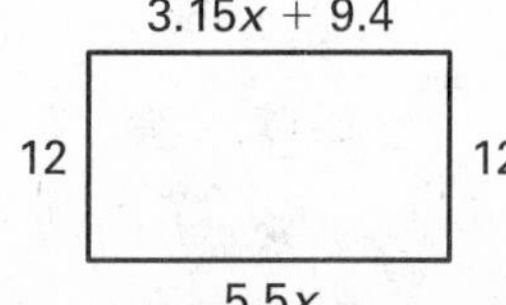

17.

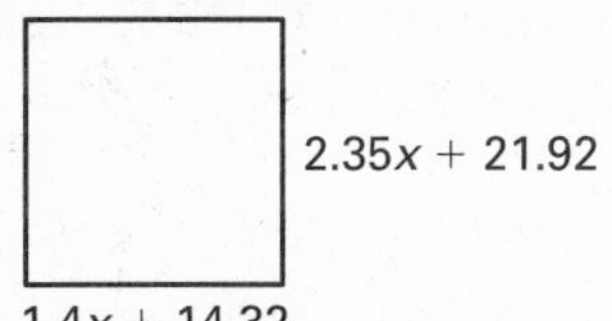

18.

$4.71x - 4.82$

5.4 | 5.4

$2.11x + 2.98$

19. The balances of two investments are changing at steady rates. One investment of \$1450 is decreasing by \$105.75 each month. The other investment of \$825 is increasing by \$144.25 each month. Write and solve an equation to find how many months it takes the balances of the investments to be the same. Let m represent the number of months.

20. A flower garden has the shape shown at the right. The diameter of the inner circle is 12 feet. Write and solve an equation to find the length x of a walkway.

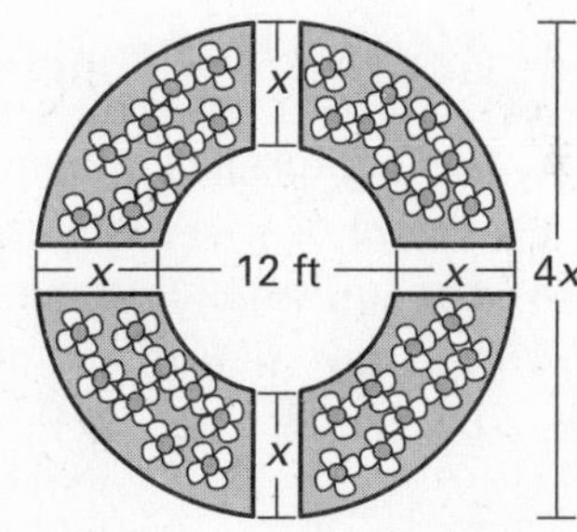

Lesson 3.3

LESSON 3.3

Name ______________________ Date ________

Study Guide

For use with pages 130–136

GOAL Solve equations with variables on both sides.

EXAMPLE 1 Solving an Equation with the Variable on Both Sides

$5n + 2 = 20n - 43$	Original equation
$5n + 2 - 5n = 20n - 43 - 5n$	Subtract $5n$ from each side.
$2 = 15n - 43$	Simplify.
$2 + 43 = 15n - 43 + 43$	Add 43 to each side.
$45 = 15n$	Simplify.
$\frac{45}{15} = \frac{15n}{15}$	Divide each side by 15.
$3 = n$	Simplify.

Answer: The solution is 3.

EXAMPLE 2 Writing and Solving an Equation

At a carnival, you spend \$6 on food and buy 12 game and ride tickets. Your friend spends nothing on food and buys 20 game and ride tickets. You both spend the same amount of money. All of the game and ride tickets cost the same amount. How much does each ticket cost?

Solution

Let c represent the cost of each ticket.

Cost of your food	+	Number of your game and ride tickets	•	Cost of each game and ride ticket	=	Number of friend's game and ride tickets	•	Cost of each game and ride ticket

$6 + 12c = 20c$	Substitute.
$6 = 8c$	Subtract $12c$ from each side and simplify.
$0.75 = c$	Divide each side by 8 and simplify.

Answer: Each game and ride ticket costs \$.75.

Exercises for Examples 1 and 2

Solve the equation. Check your solution.

1. $24z - 35 = 55 - 21z$ **2.** $9z + 12 = 6z - 30$ **3.** $5x - 19 = 20 - 8x$

4. A long-distance phone company charges \$.05 a minute, plus a monthly charge of \$5. Another long-distance phone company charges \$.09 per minute, with no monthly charge. For how many minutes per month would you have to use long distance for the phone bills from each company to be equal?

Lesson 3.3

Name ______________________________ Date __________

Study Guide

For use with pages 130–136

EXAMPLE 3 An Equation with No Solution

Solve $3(2 - x) = 5 - 3x$.

$3(2 - x) = 5 - 3x$	Write original equation.
$6 - 3x = 5 - 3x$	Distributive property

Notice that this statement is not true. The equation has no solution. As a check, you can continue solving the equation.

$6 - 3x + 3x = 5 - 3x + 3x$	Add $3x$ to each side.
$6 = 5$ ✗	Simplify.

The statement $6 = 5$ is not true, so the equation has no solution.

EXAMPLE 4 Solving an Equation with All Numbers as Solutions

$4 - 3(2t + 12) = -2 - 2(15 + 3t)$	Original equation
$4 - 6t - 36 = -2 - 30 - 6t$	Distributive property
$-6t - 32 = -6t - 32$	Simplify.

Notice that for all values of t, the statement $-6t - 32 = -6t - 32$ is true. The equation has every number as a solution.

EXAMPLE 5 Solving an Equation to Find a Perimeter

Find the perimeter of the equilateral triangle.

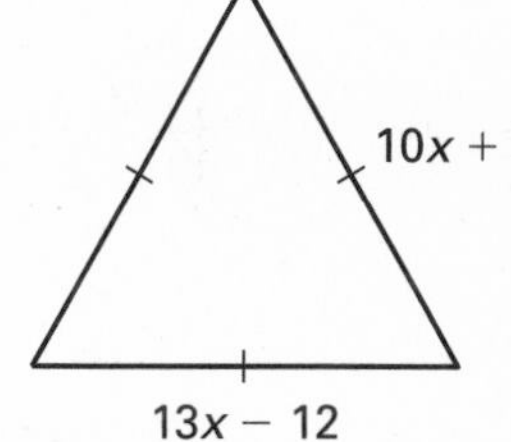

(1) An equilateral triangle has three sides of equal length. Write an equation and solve for x.

$10x + 3 = 13x - 12$	Write equation.
$3 = 3x - 12$	Subtract $10x$ from each side and simplify.
$15 = 3x$	Add 12 to each side and simplify.
$5 = x$	Divide each side by 3 and simplify.

(2) Find the length of one side by substituting 5 for x in either expression.

$10x + 3 = 10(5) + 3 = 53$

(3) To find the perimeter, multiply the length of one side by 3: $53 \cdot 3 = 159$.

Answer: The perimeter of the equilateral triangle is 159 units.

Exercises for Examples 3–5

Solve the equation. Check your solution.

5. $3(14x + 3) = 6(7x + 1) + 3$

6. $3(5 - 6z) = -14z - 2(1 + 2z) + 2$

7. Find the perimeter of a square with sides of length $9x + 11$ and $13x - 1$.

LESSON **3.3**

Name ______________________________ Date __________

Challenge Practice

For use with pages 130–136

Solve the equation. Check your solution.

1. $11 - 2x = 5(x - 5) + 15$

2. $23a + 17 = 12(5 + 4a) - 18$

3. $7(m - 5) = 15(3 - m) + 2m$

4. $14v - 3(6 - v) = 8(v - 9)$

5. Copies at the local library cost $.15 each. A new copier costs $200 and it costs $.02 to make a copy. How many copies would have to be made for the cost of making your own copies to be equal to the cost of copying at the library?

6. A smoothie at a local ice cream stand costs $2.25. You can buy a smoothie maker for $50 and buy the ingredients at a cost of $1 for each smoothie. How many smoothies would have to be made for the cost of the ice cream stand smoothies to equal the cost of the smoothies made at home?

7. Consider the equation $12x + a = 4(3x - 8)$. For what value(s) of a does the equation have a solution?

8. Consider the two figures below. For what value of x are the areas of the figures the same?

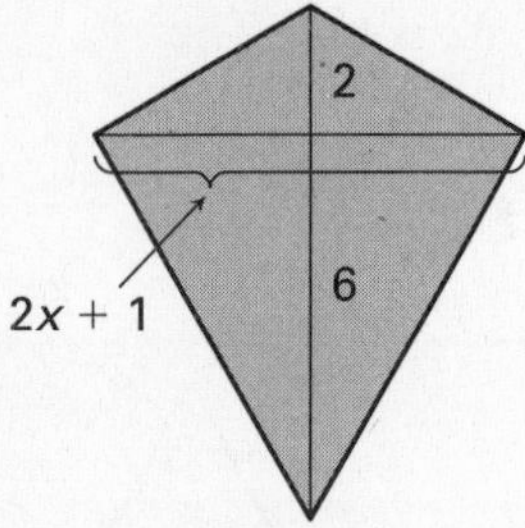

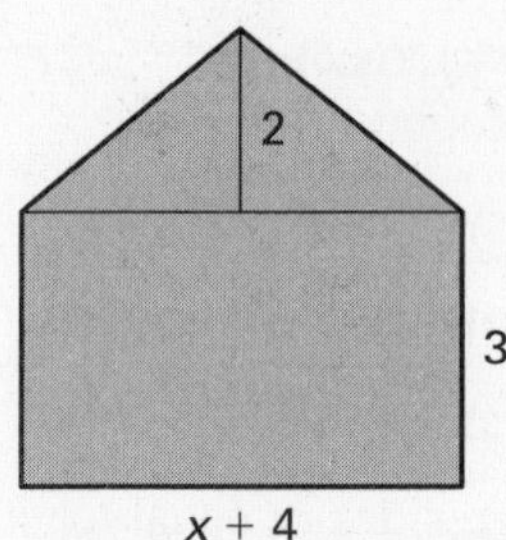

Lesson 3.3

LESSON **3.4**

Teacher's Name ______________________ Class ______ Room ______ Date ______

Lesson Plan

2-day lesson (See *Pacing and Assignment Guide*, TE page 116A)
For use with pages 138–142

GOAL **Solve inequalities using addition or subtraction.**

State/Local Objectives ____

____

✓ **Check the items you wish to use for this lesson.**

STARTING OPTIONS

____ Homework Check (3.3): TE page 134; Answer Transparencies
____ Homework Quiz (3.3): TE page 135; Transparencies
____ Warm-Up: Transparencies

TEACHING OPTIONS

____ Notetaking Guide
____ Examples: Day 1: 1, SE page 138; Day 2: 2–4, SE pages 139–140
____ Extra Examples: TE pages 139–140
____ Checkpoint Exercises: Day 1: none; Day 2: 1–4, SE page 139
____ Concept Check: TE page 140
____ Guided Practice Exercises: Day 1: 2–6, SE page 140; Day 2: 1, 7–11, SE page 140
____ Technology Activity: CRB page 37

APPLY/HOMEWORK

Homework Assignment

____ Basic: Day 1: EP p. 803 Exs. 13–15; pp. 141–142 Exs. 12–20, 33, 38, 45–49
Day 2: pp. 141–142 Exs. 21–32, 34–37, 40, 50
____ Average: Day 1: pp. 141–142 Exs. 12–20, 33, 41–43, 45–49
Day 2: pp. 141–142 Exs. 21–32, 34–40, 50
____ Advanced: Day 1: pp. 141–142 Exs. 14–20, 33, 41–48*
Day 2: pp. 141–142 Exs. 25–32, 34–40, 49, 50

Reteaching the Lesson

____ Practice: CRB pages 38–40 (Level A, Level B, Level C); Practice Workbook
____ Study Guide: CRB pages 41–42; Spanish Study Guide

Extending the Lesson

____ Real-World Problem Solving: CRB page 43
____ Challenge: SE page 142; CRB page 44

ASSESSMENT OPTIONS

____ Daily Quiz (3.4): TE page 142 or Transparencies
____ Standardized Test Practice: SE page 142

Notes ____

LESSON **3.4**

Teacher's Name ________________________ Class ______ Room ______ Date ______

Lesson Plan for Block Scheduling

1-block lesson (See *Pacing and Assignment Guide*, TE page 116A)

For use with pages 138–142

GOAL **Solve inequalities using addition or subtraction.**

State/Local Objectives ____

____

✓ Check the items you wish to use for this lesson.

Chapter Pacing Guide	
Day	**Lesson**
1	3.1; 3.2
2	3.3
3	**3.4**
4	3.5
5	3.6
6	Ch. 3 Review and Projects

STARTING OPTIONS

____ Homework Check (3.3): TE page 134; Answer Transparencies

____ Homework Quiz (3.3): TE page 135; Transparencies

____ Warm-Up: Transparencies

TEACHING OPTIONS

____ Notetaking Guide

____ Examples: 1–4, SE pages 138–140

____ Extra Examples: TE pages 139–140

____ Checkpoint Exercises: 1–4, SE page 139

____ Concept Check: TE page 140

____ Guided Practice Exercises: 1–11, SE page 140

____ Technology Activity: CRB page 37

APPLY/HOMEWORK

Homework Assignment

____ Block Schedule: pp. 141–142 Exs. 12–43, 45–50

Reteaching the Lesson

____ Practice: CRB pages 38–40 (Level A, Level B, Level C); Practice Workbook

____ Study Guide: CRB pages 41–42; Spanish Study Guide

Extending the Lesson

____ Real-World Problem Solving: CRB page 43

____ Challenge: SE page 142; CRB page 44

ASSESSMENT OPTIONS

____ Daily Quiz (3.4): TE page 142 or Transparencies

____ Standardized Test Practice: SE page 142

Notes

____

____

____

Name ____ Date ___________

LESSON **3.4**

Technology Activity

For use with pages 138–142

GOAL Graph an inequality using a graphing calculator.

EXAMPLE **Graph $x - 4 > -3$ using a graphing calculator.**

Solve and graph $x - 4 > -3$. Then check your solution using a graphing calculator.

Solution

1 Solve the inequality.

$x - 4 > -3$	Write original inequality.
$x - 4 + 4 > -3 + 4$	Add 4 to side.
$x > 1$	Simplify.

Answer: The solution is $x > 1$.

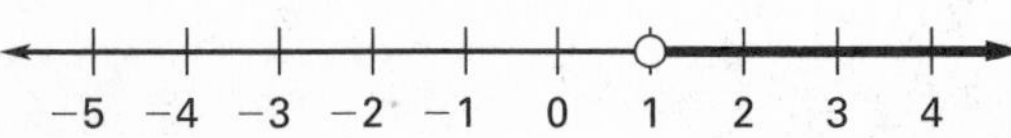

2 Now check your solution using the following keystrokes on a graphing calculator.

Keystrokes

 – 4 2nd [CATALOG] arrow up to '>'   3 GRAPH

3 Compare the graph you sketched above to the graphing calculator screen.

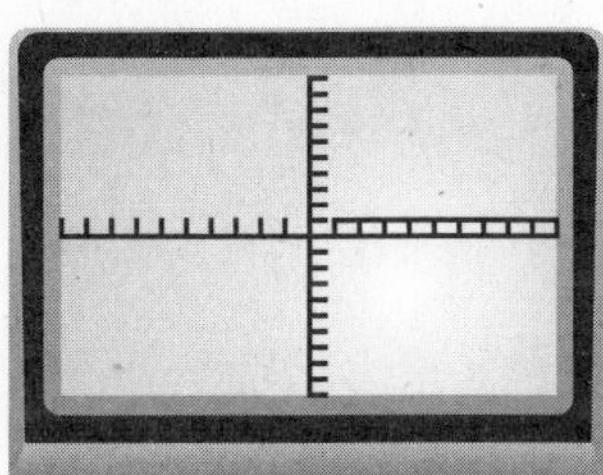

Check: Both graphs represent all numbers greater than one. Your solution checks. ✓

DRAW CONCLUSIONS **Solve and graph the inequality. Then check your solution using a graphing calculator.**

1. $x + 10 < -5$

2. $41 + x \leq 2$

3. $x - 22 > -21$

4. $x - 36 \geq 7$

5. $2.93 + x + 4 \geq 2$

6. $17.1 > 3 + x - 5.2$

7. $32.3 \geq 5 + x - 12.6$

8. $6.28 + x - 8.9 < 12$

Lesson 3.4

LESSON **3.4**

Name ______________________________ Date __________

Practice A

For use with pages 138–142

Tell whether the given number is a solution of $-8 \geq x$.

1. -8 **2.** 4 **3.** -12 **4.** 0

Write an inequality to represent the situation.

5. A boat must be at least 24 feet long.

6. A golfer's longest drive is 325 yards.

7. The maximum speed of a race car is 180 miles per hour.

8. The lowest attendance for a concert is 580.

Write an inequality represented by the graph.

9.

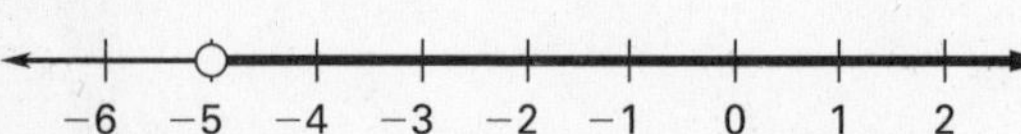

10.

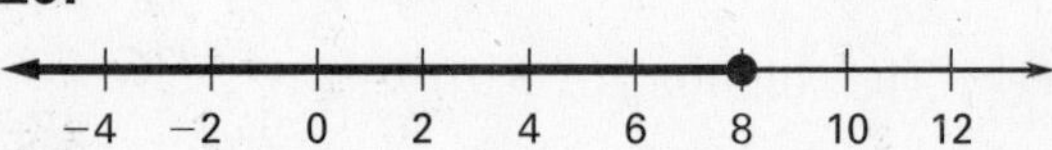

11.

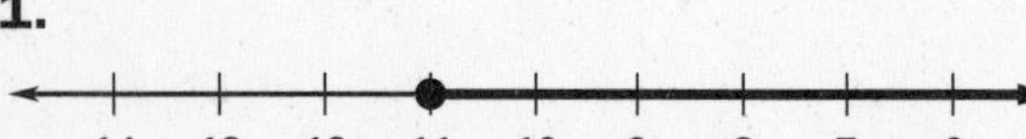

12.

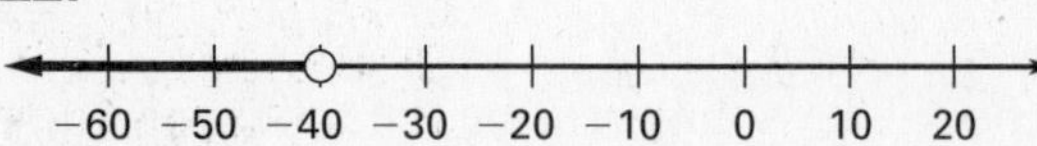

13.

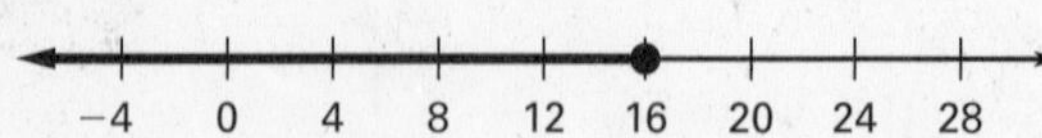

14.

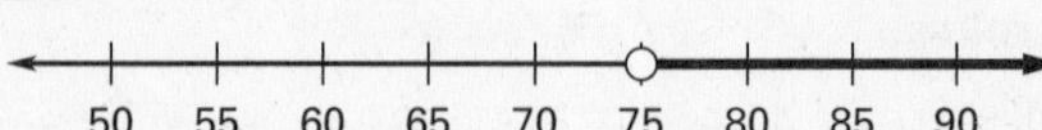

Solve the inequality. Graph your solution.

15. $x + 2 > 9$

16. $-4 + x \leq 13$

17. $-15 \geq x + 7$

18. $-1 < x - 10$

19. $x - 18 < 35$

20. $x + 24 \geq 21$

21. $x + 12 \leq -1$

22. $x - 11 > -11$

23. $30 + x < 16$

24. $-19 + x \geq -6$

25. $15 \leq x + 9$

26. $-8 > x + 8$

27. $x - 4 \geq 7.6$

28. $x + 7.1 \leq 10.3$

29. $x - 9.5 < 9.5$

30. $14.3 > x + 12.7$

31. You have \$78 in your savings account. You are buying a computer and the minimum down payment you can make is \$120. Write and solve an inequality to represent the amount of money you need to reach or exceed the minimum down payment for the computer.

Name ______________________ Date __________

LESSON **3.4**

Practice B

For use with pages 138–142

Tell whether the given number is a solution of $-8 > -17 + x - 14$.

1. -23 **2.** 23 **3.** 0 **4.** 25

Write an inequality that represents the verbal sentence.

5. Nine and four tenths plus a number is less than or equal to 14.1.

6. Thirty two plus a number minus 18 is greater than -3.

7. Six tenths plus 4.7 plus a number is greater than or equal to -5.6.

8. A number minus 6.88 is less than 22.74.

Match the inequality with the graph of its solution.

9. $x - 8 - 11 < -15$

10. $13 > -6 + 23 + x$

11. $10.45 + x - 5 > 1.45$

12. $4.5 + x - 4 > 4.5$

A. 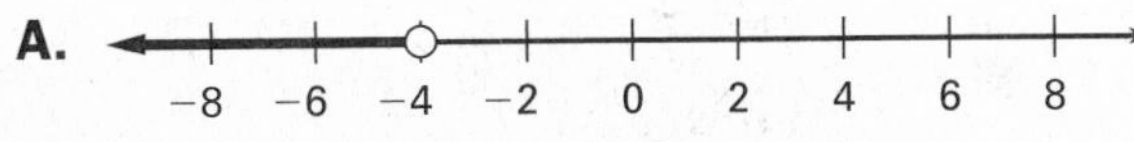

B. 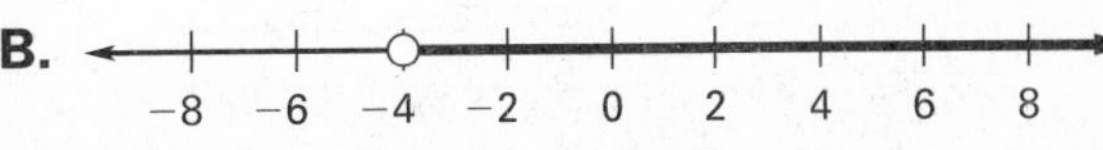

C. 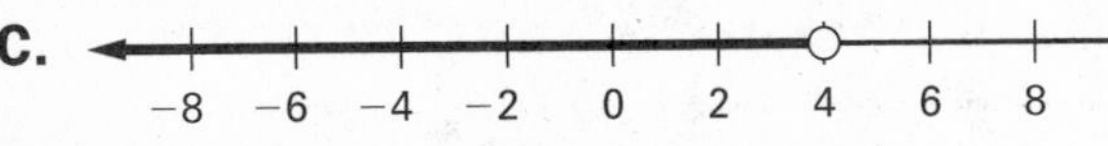

D. 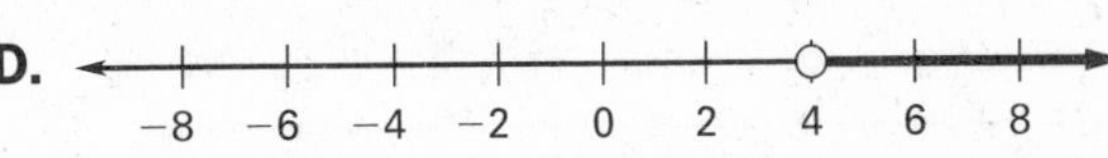

Solve the inequality. Graph your solution.

13. $7 + x + 10 < -2$

14. $5 + x - 9 \geq 4$

15. $x - 12 - 14 \leq 6$

16. $-7 - 15 + x > -15$

17. $-23 \leq x - 18 + 25$

18. $2.9 + x + 7.5 > 6$

19. $-12.1 + 16.4 + x < -3.7$

20. $-2.87 - 4.66 + x > -7.53$

21. $-1.12 \leq x + 1.53 - 4.01$

22. $10 + 11.88 + x \leq -4.5$

23. $42.76 - 21.15 \geq x + 12.9$

24. $-140.67 < 74.9 - 101.23 + x$

25. The table shows the number of preordered tickets for a three-day showing of a play. The theater has a seating capacity of 5400 people. Write and solve an inequality that represents the possible number of tickets t that can be sold at the door for each night of the play without exceeding the seating capacity of the theater.

Night	Preorder tickets
Friday	3488
Saturday	4109
Sunday	4573

26. An elevator has a weight limit of 2000 pounds. The weights in pounds of twelve people on the elevator are shown below.

175, 140, 135, 155, 170, 190, 125, 160, 150, 150, 130, 145

a. Find the total weight of the twelve people on the elevator.

b. A thirteenth person wants to get on the elevator. Write and solve an inequality that represents the weight w that person can be without exceeding the elevator's weight limit.

LESSON **3.4**

Name ______________________________ Date __________

Practice C

For use with pages 138–142

Write an inequality that represents the verbal sentence.

1. Three and five tenths plus a number is greater than or equal to -6.7.
2. Twelve and three tenths plus a number minus 5.6 is less than -2.8.
3. Negative fourteen and three tenths is less than or equal to -9.1 plus a number.
4. Seventeen and seven hundredths is greater than 31.02 plus a number minus 27.64.

Match the inequality with the graph of its solution.

5. $2.2 + x - 7.6 \geq -1.9$
6. $-6.4 < x - 9.9$
7. $16.9 \geq x + 13.4$
8. $-4.2 + x - 0.6 < -1.3$

A.

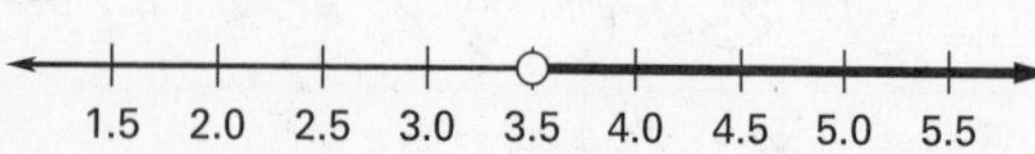

B.

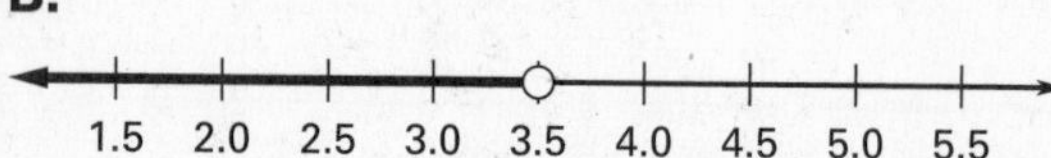

C.

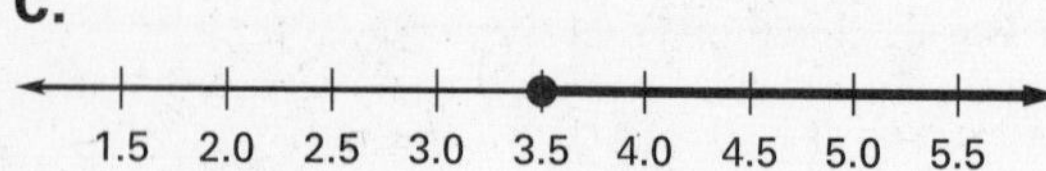

D.

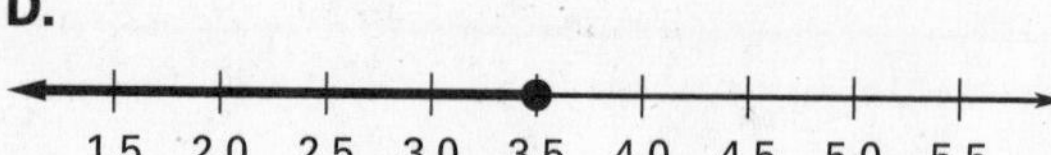

Solve the inequality. Graph your solution.

9. $12.1 + x - 2.4 < 4.9$
10. $31.5 + x + 6.7 \geq 14.8$
11. $-3.4 > x + 3.0 - 7.6$
12. $9.2 \leq -2.1 + x + 7.7$
13. $2.75 + x + 9.8 < 15.75$
14. $-2.32 + x - 2.23 \leq 4.95$
15. $7.25 \geq 4.34 + x + 2.01$
16. $-1.73 > -5.25 - 2.98 + x$
17. $-12 + 7.124 + x > -9.876$
18. $x - 0.875 - 1.004 < 8.121$
19. $x + 5.75 - 3.712 > 0.788$
20. $15.112 + x + 3.02 \leq 56.532$
21. $6.121 \leq x - 0.875 - 1.004$
22. $7.704 + x - 22.094 \leq 19.81$
23. $5.007 > 5.377 + x + 0.13$
24. $-4.123 \geq -6.75 + x + 2.027$

25. You enter a bass fishing competition. Each competitor can catch a maximum of six fish that can be entered into the competition. The six fish of the competitor in first place have a total weight of 33.5 pounds. Five of your fish have already been weighed in at 5.5 pounds, 5.75 pounds, 7 pounds, 6.5 pounds, and 5.625 pounds. Write and solve an inequality that represents the weight w your last fish needs to be in order to win first place. Then graph the inequality.

Lesson 3.4

Name ______________________________ Date __________

LESSON **3.4**

Study Guide

For use with pages 138–142

GOAL **Solve inequalities using addition or subtraction.**

> **VOCABULARY**
>
> An **inequality** is a statement formed by placing an inequality symbol between two expressions. For example, $y + 5 \le -6$ is an inequality.
>
> The **solution of an inequality** with a variable is the set of all numbers that produce true statements when substituted for the variable.
>
> **Equivalent inequalities** are inequalities that have the same solution.

EXAMPLE 1 Writing and Graphing an Inequality

Helium is the element with the lowest melting point, −272.2°C. Write an inequality that describes the melting point p (in degrees Celsius) of any other element.

Solution

Let p represent the melting point of any element. The lowest melting point is −272.2°C.

Answer: The inequality is $p \ge -272.2$. The graph is shown below.

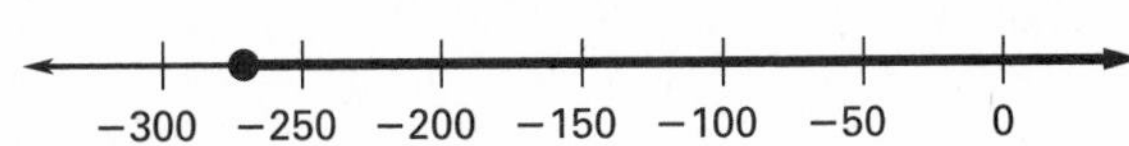

Exercises for Example 1

Write an inequality to represent the situation.

1. You need at least 85 points on the final exam to get an A in your math class.
2. You are willing to spend up to $7500 on a used car.

EXAMPLE 2 Solving an Inequality Using Subtraction

Solve $y + 11 > 7$. Graph your solution.

$y + 11 > 7$	Write original inequality.
$y + 11 - 11 > 7 - 11$	Subtract 11 from each side.
$y > -4$	Simplify.

Answer: The solution is $y > -4$.

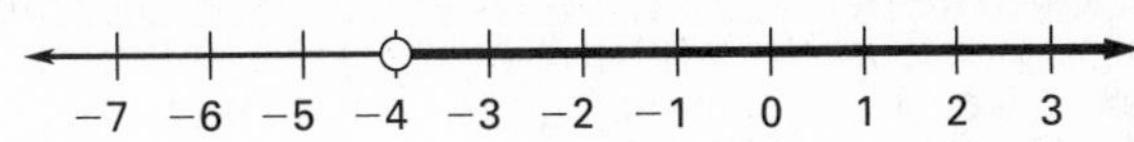

Lesson 3.4

Name ______________________________ Date __________

Study Guide

For use with pages 138–142

EXAMPLE 3 Solving an Inequality Using Addition

Solve $u - 31 < -22$. Graph and check your solution.

$u - 31 < -22$	Write original inequality.
$u - 31 + 31 < -22 + 31$	Add 31 to each side.
$u < 9$	Simplify.

Answer: The solution is $u < 9$.

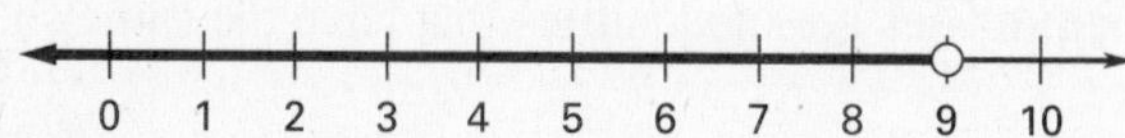

✓ **Check** Choose any number less than 9. Substitute the number into the original inequality.

$u - 31 < -22$	Write original inequality.
$0 - 31 \stackrel{?}{<} -22$	Substitute 0 for u.
$-31 < -22$ ✓	Solution checks.

Exercises for Examples 2 and 3

Solve the inequality. Graph and check your solution.

3. $y - 12 < -13$ **4.** $t + 18 > 10$ **5.** $-9 \geq m + 7$ **6.** $-3 \leq x - 6$

EXAMPLE 4 Writing and Solving an Inequality

You have 120 minutes this evening to exercise, eat dinner, and clean your room. It takes you 45 minutes to exercise and 25 minutes to eat dinner. What possible amounts of time can you spend cleaning your room?

Solution

Let t represent the time, in minutes, you spend cleaning. Write a verbal model.

Exercise time + Dinner time + Cleaning time ≤ Amount of time you have

$45 + 25 + t \leq 120$	Substitute.
$70 + t \leq 120$	Simplify.
$t \leq 50$	Subtract 70 from each side and simplify.

Answer: You can spend 50 or less minutes cleaning your room.

Exercise for Example 4

7. You owe your parents \$95. You have \$38 cash, \$20 in savings, and a job scheduled for this weekend. What possible amounts can you earn at the job in order to be able to pay your parents back in full?

Name ______________________________ Date __________

LESSON

3.4 Real-World Problem Solving

For use with pages 138–142

Vitamin A

Vitamin A is very important to the human body. It plays a key role in fending off illness both by regulating the immune system and by maintaining the skin and mucous membranes. It is important for vision because it helps maintain the surface linings of the eyes and helps us to see in the dark. Vitamin A is also important for normal bone growth. It may even reduce the risk of forming certain cancers.

While vitamin A is essential to good health, it is possible to get too much. Because it is fat soluble it can be stored in the body and very large amounts are actually toxic. The Institute of Medicine gives recommendations for safe amounts of vitamin A by age and gender. Their RDA (Recommended Dietary Allowance) for vitamin A is the minimum daily amount that is considered healthy. Their UL (Upper Intake Level) for vitamin A is the maximum daily amount considered to be safe.

In Exercises 1–6, use the following information.

Trista is looking for an affordable multivitamin to supplement her diet. She analyzed her diet to find that she consumes an average of 600 micrograms of vitamin A each day. At her age, the RDA for vitamin A is 700 micrograms and the UL is 2800 micrograms.

1. Write an inequality to represent the daily amounts of vitamin A from a multivitamin that would give her the RDA amount or more.
2. Solve the inequality in Exercise 1.
3. Write an inequality to represent the daily amounts of vitamin A from a multivitamin that would give her the UL amount or less.
4. Solve the inequality in Exercise 3.
5. Trista buys a multivitamin that has 750 micrograms of vitamin A in each capsule. Will one capsule each day give her the recommended amount of vitamin A? Explain.
6. Can Trista safely take two capsules of her multivitamin each day? Three capsules? Explain.

LESSON **3.4**

Name ______________________________ Date ________

Challenge Practice

For use with pages 138–142

Solve the inequality. Graph your solution.

1. $4.2 + t - 3.6 \leq -0.5$

2. $28 + r - 34 > 18$

3. $6.04 < 7.1 + x - 3.24$

4. $2t + 4 \geq t$

In Exercises 5 and 6, write the verbal sentence as an inequality. Then solve the inequality.

5. The sum of 8 and a number is less than -14.

6. Fourteen is greater than or equal to the difference of a number and 4.

7. At a temperature of $-78.5°C$, dry ice starts changing from a solid to a gas. Write and graph an inequality to show the temperatures at which dry ice is a solid.

8. Luann is entering a pumpkin growing contest. Last year's winning pumpkin weighed 215.5 pounds. So far, Luann's pumpkin weighs 50.5 pounds. Write and solve an inequality that represents the weight w (in pounds) that Luann's pumpkin has to gain in order to exceed the weight of last year's winning pumpkin.

9. Find all the values of r that make both of the following inequalities true: $4.2 + r \leq 7.8$ and $-6.2 + r > -10.5$. Show how you found your answer.

LESSON **3.5**

Teacher's Name ________________________ Class ______ Room ______ Date ______

Lesson Plan

2-day lesson (See *Pacing and Assignment Guide*, TE page 116A)
For use with pages 143–148

GOAL **Solve inequalities using multiplication or division.**

State/Local Objectives ____

____

✓ **Check the items you wish to use for this lesson.**

STARTING OPTIONS

____ Homework Check (3.4): TE page 141; Answer Transparencies
____ Homework Quiz (3.4): TE page 142; Transparencies
____ Warm-Up: Transparencies

TEACHING OPTIONS

____ Notetaking Guide
____ Concept Activity: SE page 143
____ Examples: Day 1: 1–2, SE pages 144–145; Day 2: 3, SE page 145
____ Extra Examples: TE page 145
____ Checkpoint Exercises: Day 1: 1–4, SE page 145; Day 2: none
____ Concept Check: TE page 145
____ Guided Practice Exercises: Day 1: 1–10, SE page 146; Day 2: 11, SE page 146
____ Technology Keystrokes for Exs. 39–44 on SE page 147: CRB page 47

APPLY/HOMEWORK

Homework Assignment

____ Basic: Day 1: pp. 146–148 Exs. 12–19, 30–33, 42–45, 50–54
Day 2: pp. 146–148 Exs. 20–27, 29, 37–41, 55–61
____ Average: Day 1: pp. 146–148 Exs. 16–19, 30–35, 42–44, 50–55
Day 2: pp. 146–148 Exs. 24–29, 36–38, 45–48, 56–61
____ Advanced: Day 1: pp. 146–148 Exs. 16–19, 32–35, 42–46, 54–58
Day 2: pp. 146–148 Exs. 24–29, 36–38, 47–51*, 59–61

Reteaching the Lesson

____ Practice: CRB pages 48–50 (Level A, Level B, Level C); Practice Workbook
____ Study Guide: CRB pages 51–52; Spanish Study Guide

Extending the Lesson

____ Challenge: SE page 148; CRB page 53

ASSESSMENT OPTIONS

____ Daily Quiz (3.5): TE page 148 or Transparencies
____ Standardized Test Practice: SE page 148

Notes ____

LESSON **3.5**

Teacher's Name ______________________ Class ______ Room ______ Date ______

Lesson Plan for Block Scheduling

1-block lesson (See *Pacing and Assignment Guide*, TE page 116A)
For use with pages 143–148

GOAL **Solve inequalities using multiplication or division.**

State/Local Objectives ____

____

✓ **Check the items you wish to use for this lesson.**

Chapter Pacing Guide	
Day	**Lesson**
1	3.1; 3.2
2	3.3
3	3.4
4	**3.5**
5	3.6
6	Ch. 3 Review and Projects

STARTING OPTIONS

____ Homework Check (3.4): TE page 141; Answer Transparencies
____ Homework Quiz (3.4): TE page 142; Transparencies
____ Warm-Up: Transparencies

TEACHING OPTIONS

____ Notetaking Guide
____ Concept Activity: SE page 143
____ Examples: 1–3, SE pages 144–145
____ Extra Examples: TE page 145
____ Checkpoint Exercises: 1–4, SE page 145
____ Concept Check: TE page 145
____ Guided Practice Exercises: 1–11, SE page 146
____ Technology Keystrokes for Exs. 39–44 on SE page 147: CRB page 47

APPLY/HOMEWORK

Homework Assignment

____ Block Schedule: pp. 146–148 Exs. 16–19, 24–38, 42–48, 50–61

Reteaching the Lesson

____ Practice: CRB pages 48–50 (Level A, Level B, Level C); Practice Workbook
____ Study Guide: CRB pages 51–52; Spanish Study Guide

Extending the Lesson

____ Challenge: SE page 148; CRB page 53

ASSESSMENT OPTIONS

____ Daily Quiz (3.5): TE page 148 or Transparencies
____ Standardized Test Practice: SE page 148

Notes

____

____

____

LESSON **3.5**

Name ______________________ Date __________

Technology Keystrokes

For use with Exercises 39–44, page 147

TI-34 II

39. 40.94 [÷] [(−)] 8.9 [ENTER =]

40. 8.5 [×] 2.4 [ENTER =]

41. [(−)] 3.4 [×] 7.2 [ENTER =]

42. 10.71 [÷] 6.3 [ENTER =]

43. 43.68 [÷] [(−)] 3.9 [ENTER =]

44. 6.5 [×] [(−)] 9.1 [ENTER =]

TI-73 Explorer

39. 40.94 [÷] [(−)] 8.9 [ENTER]

Graph: [Y=] [x] [2nd] [CATALOG] Arrow up to '≤'. [ENTER] [(−)] 4.6

[GRAPH] Adjust window if necessary.

40. 8.5 [×] 2.4 [ENTER]

Graph: [Y=] [x] [2nd] [CATALOG] Arrow up to '≥'. [ENTER] 20.4

[GRAPH] Adjust window if necessary.

41. [(−)] 3.4 [×] 7.2 [ENTER]

Graph: [Y=] [x] [2nd] [CATALOG] Arrow up to '<'. [ENTER] [(−)] 24.48

[GRAPH] Adjust window if necessary.

42. 10.71 [÷] 6.3 [ENTER]

Graph: [Y=] [x] [2nd] [CATALOG] Arrow up to '>'. [ENTER] 1.7

[GRAPH] Adjust window if necessary.

43. 43.68 [÷] [(−)] 3.9 [ENTER]

Graph: [Y=] [x] [2nd] [CATALOG] Arrow up to '≥'. [ENTER] [(−)] 11.2

[GRAPH] Adjust window if necessary.

44. 6.5 [×] [(−)] 9.1 [ENTER]

Graph: [Y=] [x] [2nd] [CATALOG] Arrow up to '≥'. [ENTER] [(−)] 59.15

[GRAPH] Adjust window if necessary.

Name ______________________________ Date __________

LESSON **3.5**

Practice A

For use with pages 143–148

Lesson 3.5

Tell whether the given number is a solution of $\frac{x}{8} \le -3$.

1. -16 **2.** -24 **3.** 0 **4.** -28

Match the inequality with the graph of its solution.

5. $\frac{x}{-2} \ge 26$ **A.** –15 –14 –13 –12 –11 –10

6. $-2x \ge 26$ **B.** –56 –54 –52 –50 –48 –46

7. $\frac{x}{-2} \le 26$ **C.** –56 –54 –52 –50 –48 –46

8. $-2x \le 26$ **D.** –15 –14 –13 –12 –11 –10

Solve the inequality. Graph your solution.

9. $\frac{x}{3} > -2$ **10.** $\frac{x}{8} \le 8$ **11.** $4x \ge -28$ **12.** $15x < 45$

13. $2x > -34$ **14.** $3x \ge 33$ **15.** $\frac{x}{9} < 6$ **16.** $\frac{x}{-11} \ge -11$

17. $\frac{x}{10} \ge -1$ **18.** $\frac{x}{-5} < 12$ **19.** $-14x < 84$ **20.** $-5x \le -45$

21. $-6x \ge 48$ **22.** $-20x > -100$ **23.** $\frac{x}{-2} > -7$ **24.** $\frac{x}{-13} \le -4$

25. You want to buy some new DVDs and spend less than $75. A store advertises a sale where all DVDs are $15. Write and solve an inequality to find the number of DVDs *d* you can buy.

26. Tickets for a basketball tournament cost $3. The tournament wants to make $1575 the first night in ticket sales. Write and solve an inequality to find the number of tickets *t* that has to be sold to make at least $1575.

27. You get a part-time job delivering flowers for a flor... $2.50 for each delivery. Write and solve an inequal... deliveries *d* you need to make in order to earn at le...

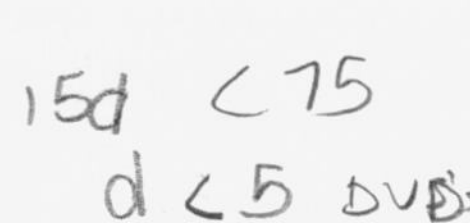

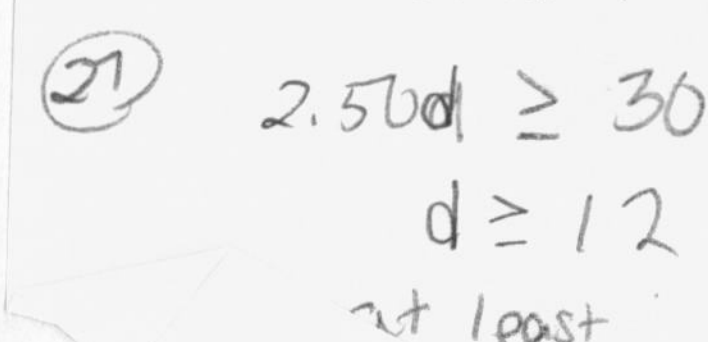

Name ______________________ Date ______________

LESSON 3.5

Practice B

For use with pages 143–148

Tell whether the given number is a solution of $-1.5x > -12$.

1. -4 **2.** 12 **3.** 0 **4.** 8

Match the inequality with the graph of its solution.

5. $\frac{x}{1.4} > -10$

6. $-3.5x > 14$

7. $\frac{x}{-1.4} > 10$

8. $-3.5x < 14$

A. 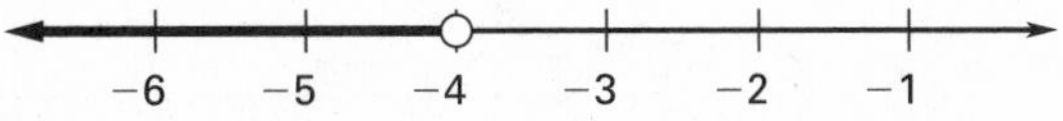

B. 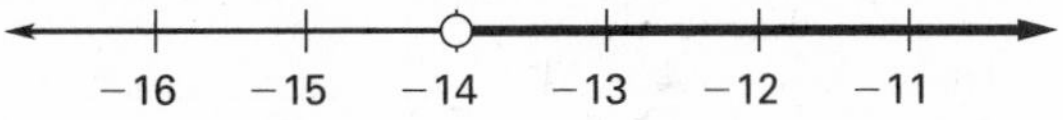

C. 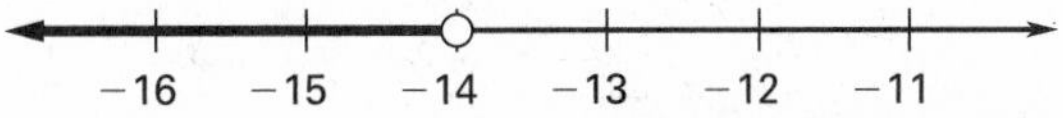

D. 

Solve the inequality. Graph your solution.

9. $\frac{x}{-10} \le 22$ **10.** $\frac{x}{25} > -30$ **11.** $-13x < -208$ **12.** $45x \le -855$

13. $1.6x \le -11.2$ **14.** $-5.3x > 21.2$ **15.** $-10.7 > \frac{x}{-4}$ **16.** $8.3 \le \frac{x}{-5}$

17. $\frac{x}{1.3} \ge 7.1$ **18.** $\frac{x}{-5.6} < 2.8$ **19.** $-3.8x \ge 28.5$ **20.** $10.4x > 520$

21. $-0.1x \le -2.5$ **22.** $-9.8x \ge 44.1$ **23.** $\frac{x}{-12.7} \ge -2.2$ **24.** $\frac{x}{4.2} < -20.45$

Write the verbal sentence as an inequality. Then solve the inequality.

25. A number divided by 3.5 is greater than or equal to 7.8.

26. The product of a number and -5 is less than -1.6.

27. The product of a number and -0.9 is greater than 27.

28. A number divided by -4.75 is greater than or equal to -20.

29. You need to complete at least 300 math problems in 4 days for a homework assignment. How many exercises should you complete each day?

30. An admission pass for an art museum is \$4.50. Write and solve an inequality to find the number of passes p that must be sold for the museum to make at least \$7200.

Name ______________________ Date __________

LESSON

3.5 Practice C

For use with pages 143–148

Lesson 3.5

Match the inequality with the graph of its solution.

1. $\frac{x}{-6.25} > -4.8$

2. $\frac{x}{6.25} < -4.8$

3. $\frac{x}{-6.25} < 4.8$

4. $\frac{x}{6.25} > 4.8$

A. 28 29 30 31 32 33

B. −32 −31 −30 −29 −28 −27

C. 28 29 30 31 32 33

D. −32 −31 −30 −29 −28 −27

Solve the inequality. Graph your solution.

5. $\frac{x}{0.45} \geq -16.2$

6. $\frac{x}{2.96} < 11.5$

7. $1.875x > 105$

8. $-3.01x \leq -48.16$

9. $-12.9x \geq 103.2$

10. $-0.15x > -9.72$

11. $\frac{x}{-9.4} > 60$

12. $\frac{x}{-16.25} < -8.4$

13. $\frac{x}{-3.8} > -1.75$

14. $\frac{x}{-1.14} \leq 150$

15. $25.8x \leq -12.9$

16. $31.8 < -42.4x$

17. $-0.008x \leq -0.336$

18. $\frac{x}{8.125} > 4000$

19. $\frac{x}{-0.001} \geq 370$

20. $-14.625x \geq -175.5$

Use the given area of each rectangle to write and solve an inequality that represents the possible lengths of the rectangle.

21. Area < 14.85 cm^2

2.75 cm

ℓ

22. Area ≤ 32.87 ft^2

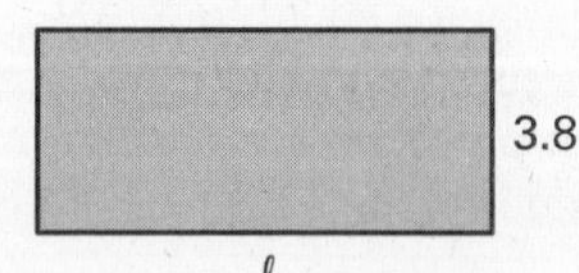

3.8 ft

ℓ

23. Area > 67.5 yd^2

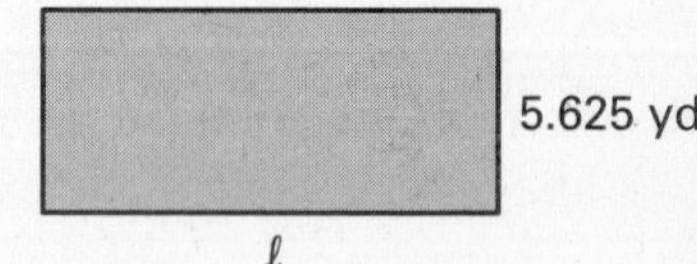

5.625 yd

ℓ

Write an inequality that is equivalent to the given inequality.

24. $4x \leq 10$

25. $-8x > 7$

26. $-0.5x < -12$

27. $1.3x \geq -11.6$

LESSON **3.5**

Name ______________________________ Date __________

Study Guide

For use with pages 143–148

Lesson 3.5

GOAL **Solve inequalities using multiplication or division.**

EXAMPLE 1 Solving an Inequality Using Multiplication

Solve $\frac{y}{17} \le -2$. Graph your solution.

$\frac{y}{17} \le -2$	Write original inequality.
$17 \cdot \frac{y}{17} \le 17(-2)$	Multiply each side by 17.
$y \le -34$	Simplify.

Answer: The solution is $y \le -34$.

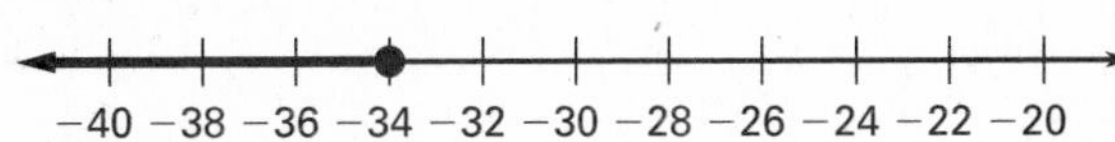

EXAMPLE 2 Solving an Inequality Using Division

Solve $-8x < -104$. Graph your solution.

$-8x < -104$	Write original inequality.
$\frac{-8x}{-8} > \frac{-104}{-8}$	Divide each side by -8. Reverse inequality symbol.
$x > 13$	Simplify.

Answer: The solution is $x > 13$.

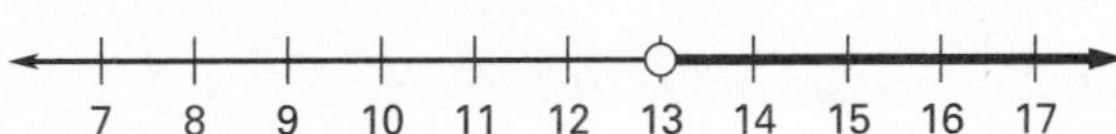

Exercises for Examples 1 and 2

Solve the inequality. Graph your solution.

1. $\frac{h}{6} < -4$ **2.** $5u > -35$ **3.** $-7y \ge -63$ **4.** $18 \ge \frac{x}{-3}$

Lesson 3.5

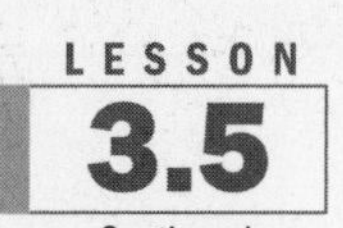

LESSON 3.5 Continued

Name ____________________ Date __________

Study Guide

For use with pages 143–148

EXAMPLE 3 Writing and Solving an Inequality

If you are at-bat 250 times this baseball season, how many hits must you get to have a batting average of at least 0.452?

Solution

Let h represent the number of hits. Write a verbal model.

$$\frac{\text{Hits}}{\text{At-bats}} \geq \text{Target batting average}$$

$\frac{h}{250} \geq 0.452$	Substitute.
$250 \cdot \frac{h}{250} \geq 250 \cdot 0.452$	Multiply each side by 250.
$h \geq 113$	Simplify.

Answer: You have to get at least 113 hits to achieve a batting average of at least 0.452.

Exercise for Example 3

5. You earn $6 per hour at your after-school job. How many hours must you work this week to earn at least $72?

LESSON **3.5**

Name ____ Date ____________

Challenge Practice

For use with pages 143–148

Solve the inequality. Graph the solution.

1. $-6.2x \le -31$

2. $6.5 > \frac{a}{8}$

3. $-2.1 < \frac{n}{4.8}$

4. $\frac{w}{8.5} \ge -16.2$

5. Elisa is taking a 225-mile trip and she wants to reach her destination in less than $4\frac{1}{2}$ hours. What average speeds allow Elisa to reach her goal if she does not make any stops?

6. Your art class is selling hand-made greeting cards to raise money for a trip to an art museum. Your class makes $1.25 for each card you sell. How many cards must your class sell to make at least $250?

7. The length of a rectangle is two times its width. If the perimeter of the rectangle can be no more than 54 units, what are the possible widths of the rectangle?

8. Find all the values of x that make both of the following inequalities true: $\frac{x}{3.5} < 6$ and $4x \ge -16$. Show how you found your answer.

9. Find all the values of x that make both of the following inequalities true: $5.5x < 11$ and $-2.4x \ge -12$. Show how you found your answer.

LESSON **3.6**

Teacher's Name ____________ Class ______ Room ______ Date ______

Lesson Plan

2-day lesson (See *Pacing and Assignment Guide*, TE page 116A)

For use with pages 149–153

GOAL **Solve multi-step inequalities.**

State/Local Objectives ____________

Lesson 3.6

✓ **Check the items you wish to use for this lesson.**

STARTING OPTIONS

_____ Homework Check (3.5): TE page 146; Answer Transparencies

_____ Homework Quiz (3.5): TE page 148; Transparencies

_____ Warm-Up: Transparencies

TEACHING OPTIONS

_____ Notetaking Guide

_____ Examples: Day 1: 2, SE page 150; Day 2: 1, 3, SE pages 149–150

_____ Extra Examples: TE page 150

_____ Checkpoint Exercises: Day 1: none; Day 2: 1, SE page 149

_____ Concept Check: TE page 150

_____ Guided Practice Exercises: Day 1: 3–8, SE page 151; Day 2: 1–2, 9 SE page 151

APPLY/HOMEWORK

Homework Assignment

_____ Basic: Day 1: SRH p. 802 Exs. 1–3; pp. 151–153 Exs. 10–17, 25–30, 47–49
Day 2: pp. 151–153 Exs. 18–23, 31, 35–46

_____ Average: Day 1: pp. 151–153 Exs. 10–17, 25–27, 34–43
Day 2: pp. 151–153 Exs. 18–24, 28–32, 44–49

_____ Advanced: Day 1: pp. 151–153 Exs. 12–17, 25–27, 39–46
Day 2: pp. 151–153 Exs. 18–24, 28–34*, 47–49

Reteaching the Lesson

_____ Practice: CRB pages 56–58 (Level A, Level B, Level C); Practice Workbook

_____ Study Guide: CRB pages 59–60; Spanish Study Guide

Extending the Lesson

_____ Challenge: SE page 152; CRB page 61

ASSESSMENT OPTIONS

_____ Daily Quiz (3.6): TE page 153 or Transparencies

_____ Standardized Test Practice: SE page 153

_____ Quiz (3.4–3.6): Assessment Book page 30

Notes ____________

LESSON **3.6**

Teacher's Name ______________________ Class ______ Room ______ Date ______

Lesson Plan for Block Scheduling

1-block lesson (See *Pacing and Assignment Guide*, TE page 116A)

For use with pages 149–153

GOAL **Solve multi-step inequalities.**

State/Local Objectives ____

____

✓ **Check the items you wish to use for this lesson.**

Chapter Pacing Guide	
Day	**Lesson**
1	3.1; 3.2
2	3.3
3	3.4
4	3.5
5	**3.6**
6	Ch. 3 Review and Projects

STARTING OPTIONS

_____ Homework Check (3.5): TE page 146; Answer Transparencies

_____ Homework Quiz (3.5): TE page 148; Transparencies

_____ Warm-Up: Transparencies

TEACHING OPTIONS

_____ Notetaking Guide

_____ Examples: 1–3, SE pages 149–150

_____ Extra Examples: TE page 150

_____ Checkpoint Exercises: 1, SE page 149

_____ Concept Check: TE page 150

_____ Guided Practice Exercises: 1–9, SE page 151

APPLY/HOMEWORK

Homework Assignment

_____ Block Schedule: pp. 151–153 Exs. 10–15, 18–32, 34–49

Reteaching the Lesson

_____ Practice: CRB pages 56–58 (Level A, Level B, Level C); Practice Workbook

_____ Study Guide: CRB pages 59–60; Spanish Study Guide

Extending the Lesson

_____ Challenge: SE page 152; CRB page 61

ASSESSMENT OPTIONS

_____ Daily Quiz (3.6): TE page 153 or Transparencies

_____ Standardized Test Practice: SE page 153

_____ Quiz (3.4–3.6): Assessment Book page 30

Notes

____

____

____

____

Name ______________________ Date ________

LESSON **3.6**

Practice A

For use with pages 149–153

Tell whether the given number is a solution of $\frac{x}{-4} + 5 > 9$.

1. -4 **2.** -20 **3.** -16 **4.** 4

Tell whether the given number is a solution of $7x - 6 \le 4x + 9$.

5. 8 **6.** 5 **7.** 0 **8.** -5

Match the inequality with the graph of its solution.

9. $\frac{x}{3} - 7 \le -5$

10. $-5x + 6 \le -24$

11. $\frac{x}{-3} + 9 \le 11$

12. $-5x - 6 \ge 24$

A. 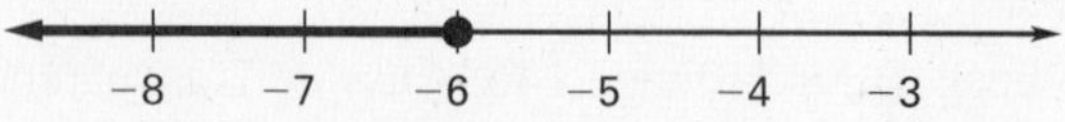

B. 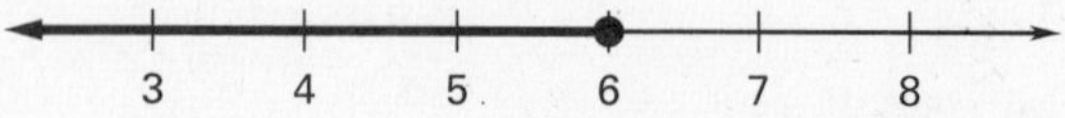

C. 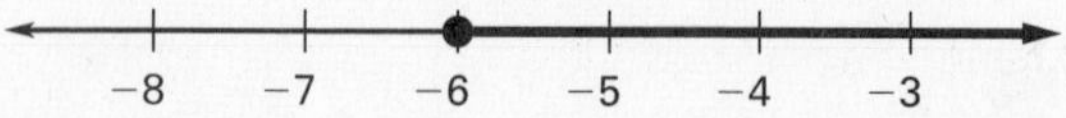

D. 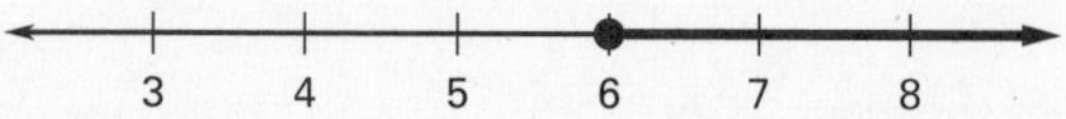

Solve the inequality. Graph your solution.

13. $3x + 8 < 8$ **14.** $4x - 7 \ge 5$ **15.** $13 - 2x > -3$

16. $-1 + 5x \le -26$ **17.** $\frac{x}{2} + 5 > 8$ **18.** $\frac{x}{4} - 6 \le -10$

19. $-11 + \frac{x}{7} < -14$ **20.** $\frac{x}{-3} - 1 \le 11$ **21.** $16 \le \frac{x}{20} - 13$

22. $5x + 12 > 3x - 8$ **23.** $10x - 6 \le -x + 38$ **24.** $-6x - 1 < -2x + 7$

25. $6x + 7 \le x + 32$ **26.** $8 + x > 2x - 9$ **27.** $13x - 8 \ge -2x + 97$

28. Your school's basketball team is trying to break the school record for points scored in a season. Your team has already scored 736 points this season. The record is 1076 points. With 10 games remaining on the schedule, how many points per game does your team need to average to break the record? Use the verbal model below to write and solve an inequality to solve the problem. Let p represent the points scored per game.

Points scored this season	+	Number of games left	·	Points scored per game	>	School record

LESSON **3.6**

Name ______________________________ Date __________

Practice B

For use with pages 149–153

Tell whether the given number is a solution of $2(3x + 1) \geq 7x + 4$.

1. -1 **2.** -2 **3.** -10 **4.** 0

Match the inequality with the graph of its solution.

5. $3(4x - 1) \leq 10x + 25$

6. $2(14 - 3x) \leq -4x$

7. $-7x + 17 \geq 115$

8. $\frac{x + 4}{5} \geq -2$

A. −16 −15 −14 −13 −12 −11

B. 12 13 14 15 16 17

C. −16 −15 −14 −13 −12 −11

D. 12 13 14 15 16 17

Solve the inequality. Graph your solution.

9. $-6x - 15 > 57$

10. $22 > \frac{x}{-12} + 4$

11. $-3(2 - x) \leq 2x - 9$

12. $6(5 - 2x) < 5x + 13$

13. $\frac{3x - 1}{4} < 8$

14. $\frac{2x + 5}{3} \geq -7$

15. $\frac{-x - 11}{3} \leq 21$

16. $-8 < \frac{5x + 4}{7}$

17. $4x + 22 > -2(14 + 3x)$

18. $-4(x + 10) \geq -7x + 65$

19. $8(3x - 19) < 15x + 73$

20. $74 < \frac{-17x + 30}{5}$

21. $\frac{25x - 41}{13} \leq 18$

22. $12(2x - 13) > 117 - 15x$

23. $29x - 515 \leq -14(8 - 3x)$

24. The golf course you play at charges \$22 per round of golf. You can either rent golf clubs at the course for \$8 or you can buy your own set of clubs for \$160. Write and solve an inequality to find the number of rounds of golf you need to play in order for the cost of purchasing clubs to be less than the cost of renting clubs. Let r represent the number of rounds of golf.

25. For what values of x is the area of the rectangle shown greater than 100 square units?

26. For what values of x is the perimeter of the rectangle shown greater than 50 units?

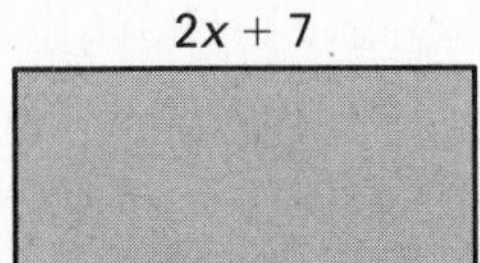

Name ______________________ Date __________

LESSON **3.6**

Practice C

For use with pages 149–153

Lesson 3.6

Tell whether the given number is a solution of $\frac{x-6}{-1.4} \geq 15$.

1. -18 **2.** -15 **3.** -14 **4.** -2

Match the inequality with the graph of its solution.

5. $-8.6x + 13 + 4.6x \leq 53$

6. $\frac{x+6}{3.2} \leq 5$

7. $\frac{-6.8x - 7.1}{2.9} \geq 21$

8. $14.5(6 - x) \leq -9.5x + 37$

A.
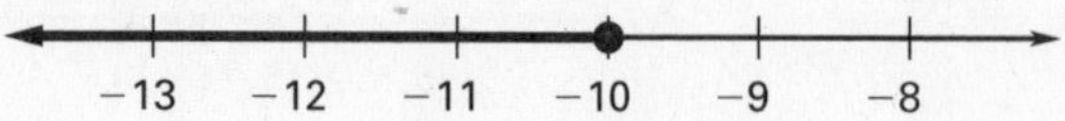

B.
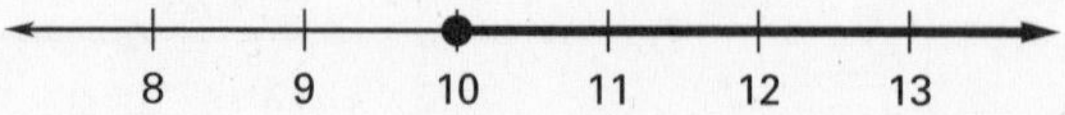

C.
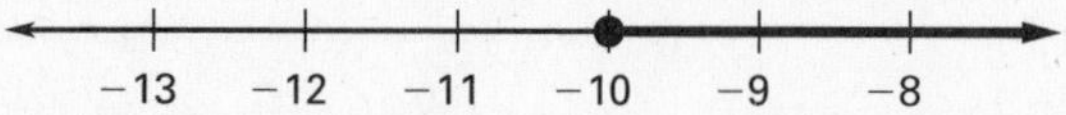

D.
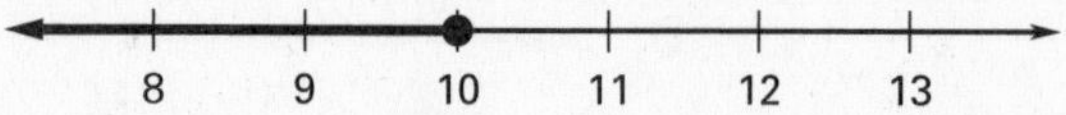

Solve the inequality. Graph your solution.

9. $-3.5x - 13 \leq 32.5$

10. $\frac{x}{6.4} + 9 > -11$

11. $-4(7.4 - x) < 3x + 9.6$

12. $0.5(11.8 - x) \geq -x - 8.1$

13. $\frac{-1.2x + 4.8}{10} < -3$

14. $\frac{13.2 + 2x}{-5.5} \geq 0.4$

15. $\frac{-6x - 2.5}{5} > -8$

16. $17 < \frac{12.5x + 8.5}{-2}$

17. $8x + 11.4 > -3(4.9 - 3x)$

18. $-0.25(5x + 16) \leq -7.25x + 32$

19. $120 > \frac{-x - 46.8}{0.2}$

20. $\frac{6x + 84.3}{-7.5} \leq -18.2$

21. $7(0.1x - 15) < 95 + 0.45x$

22. $3.38x - 9.8 \geq 2.7(8 - 3.4x)$

For Exercises 23 and 24, write and solve an inequality.

23. Anita works part-time as a waitress. She earns \$6.95 per hour. For the week, she makes \$42.90 in tips. How many hours does she need to work to earn more than \$168 for the week?

24. You are taking a taxi. The driver charges an initial fare of \$2, plus \$1.75 for every mile driven. How far can you travel in the taxi if you want to leave the driver a \$3.50 tip, and you want to spend no more than \$16?

LESSON **3.6**

Name ______________________________ Date __________

Study Guide

For use with pages 149–153

GOAL Solve multi-step inequalities.

EXAMPLE 1 Writing and Solving a Multi-Step Inequality

You are organizing a trip to a baseball game. Tickets are $12 per person, and the cost to rent the bus will be divided evenly. Find the possible costs of the bus rental to keep the cost per person under $20 if 30 people sign up to go on the trip.

Solution

Let t represent the total cost of the bus rental. Write a verbal model.

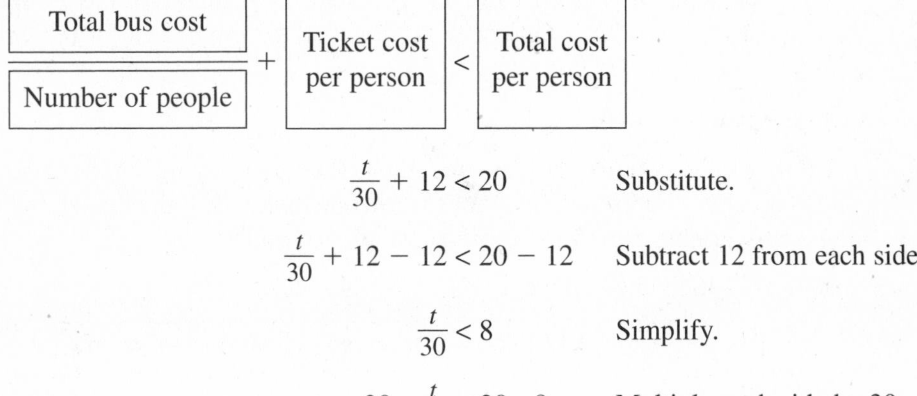

$\frac{t}{30} + 12 < 20$ Substitute.

$\frac{t}{30} + 12 - 12 < 20 - 12$ Subtract 12 from each side.

$\frac{t}{30} < 8$ Simplify.

$30 \cdot \frac{t}{30} < 30 \cdot 8$ Multiply each side by 30.

$t < 240$ Simplify.

Answer: The total cost to rent the bus must be less than $240 to keep the cost per person under $20.

Exercises for Example 1

1. You are collecting sponsors for a 10-mile walk-a-thon. So far, you have collected $230 in donations. How much must the last sponsor pledge per mile to reach or exceed your goal of $300?
2. You are biking at a rate of 30 miles per hour. You have already biked 20 miles. How many more hours must you bike to surpass your goal of 50 miles?

EXAMPLE 2 Solving a Multi-Step Inequality

$-x + 31 < 19$ Original inequality

$-x + 31 - 31 < 19 - 31$ Subtract 31 from each side.

$-x < -12$ Simplify.

$\frac{-x}{-1} > \frac{-12}{-1}$ Divide each side by -1. Reverse inequality symbol.

$x > 12$ Simplify.

Lesson 3.6

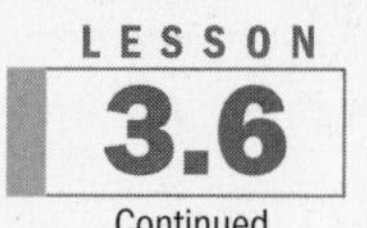

Name ______________________ Date ________

Study Guide

For use with pages 149–153

Lesson 3.6

Exercises for Example 2

Solve the inequality. Then graph the solution.

3. $\frac{x}{-8} + 3 \leq -12$ **4.** $\frac{m}{2} - 7 < -3$ **5.** $-13x + 11 \geq 180$

EXAMPLE 3 Combining Like Terms in a Multi-Step Inequality

You are going to a dinner and a movie with a group of people. Individual dinners are $7 per person, or the group can pay a lump sum of $105 for a buffet. Tickets to the movie are $5 each. How many people have to attend for the group cost of the buffet dinner and a movie to be less than the group cost for individual dinners and a movie?

Solution

There are two options: buying individual dinners or buying a buffet for everyone to share. Let p represent the number of people that attend the dinner and movie. Write a variable expression for the cost of each option.

Option 1: Individual Dinners

(Dinner price + Movie ticket price) • Number of people ⟶ $12p$

Option 2: Buffet

Buffet price + Movie ticket price • Number of people ⟶ $105 + 5p$

To find the values of p for which the group cost of option 2 is less than the group cost of option 1, write and solve an inequality.

Cost of option 2 < Cost of option 1

$105 + 5p < 12p$	Substitute.
$105 < 7p$	Subtract $5p$ from each side and simplify.
$15 < p$	Divide each side by 7 and simplify.

Answer: More than 15 people have to attend for the group buffet and movie option to be less than the individual dinner and movie option.

Exercise for Example 3

6. Tickets to your favorite team's games are $12 each, and season tickets are $396 for the same type of seat. Parking is $5 per game. How many times do you have to use the season pass for the total cost of the season ticket option to be less than the total cost of the individual-game ticket option?

LESSON **3.6**

Name ______________________________ Date __________

Challenge Practice

For use with pages 149–153

Solve the inequality. Graph your solution.

1. $3b - 6 > 7(b + 2)$

2. $\frac{6 - x}{4} \geq 1$

3. $\frac{2x + 4}{6} < -6$

4. $\frac{3p - 2}{-3} \leq 8$

5. A video game system costs \$185 and one video game costs \$14.95. You can spend no more than \$280 on the system and games. What is the greatest number of games you can buy?

6. Mohan can spend no more than \$40 on school supplies. He decides to buy a ream of paper for \$3.98 and spend the rest of his money on 3-ring binders that cost \$3.75 each. How many binders can Mohan buy?

7. At a local lake, there are two boat rental stores. Store A requires a \$50 deposit and charges \$35 per hour for the boat rental. Store B charges \$55 per hour and no deposit. How long do you have to rent a boat before the cost at store A is less than the cost at store B?

8. Use the figures below. What does x have to be in order for the area of the triangle to be greater than the area of the rectangle?

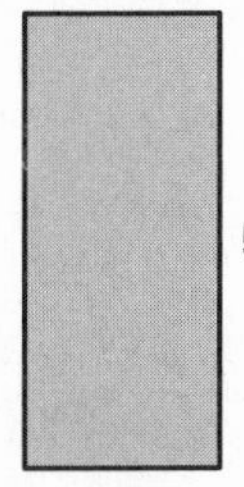

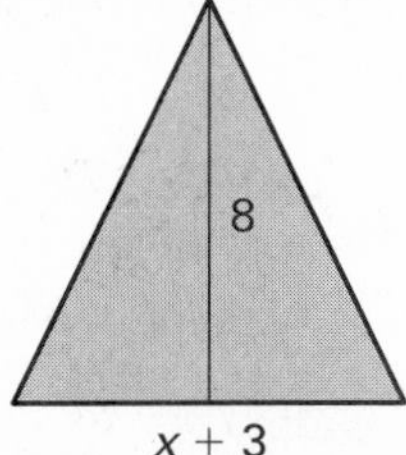

9. The length ℓ of a rectangle is 5 units longer than its width w. If the perimeter of the rectangle must be at least 34 units, what are the possible lengths of the rectangle?

Lesson 3.6

CHAPTER 3

Name ______________________________ Date __________

Chapter Review Games and Activities

For use after Chapter 3

Cross Equation Puzzle

Solve for x to complete the cross equation puzzle.

1	2	3		4	5
	6		7		
8			9		
10		11			12
				13	
	14				

Across

1. $4x - 148 = 2x + 566$

4. $3(x + 1) = 48$

6. $\frac{x}{4} - 138 = 89$

9. $7(x - 13) = 56$

10. $5x - 3099 = 4027 - 2x$

13. $4(x + 19) = 212$

14. $2x - 8 = 3x - 467$

Down

2. $2x - 83 = 35$

3. $3(2x - 55) = 255$

5. $212 - 3x = 53$

7. $\frac{x}{4} + 135 = 342$

8. $1506 - 2x = 3x + 646 - x$

11. $8x - 153 - 3x = 237 + 3x$

12. $\frac{x}{7} + 13 = 62$

CHAPTER 3

Name ______________________ Date ________

Real-Life Project: School Play

For use after Chapter 3

Objective **Analyze the profitability of a school play.**

Materials pencil, paper, access to the Internet

Investigation *Getting Going* Your school has an annual play to raise money. You are the supervisor of a committee to organize the play. You decide to have three nights of performances, on Friday, Saturday, and Sunday. The income depends on the number of people that attend the play and pay the admission price. The profit that the school makes is the income minus the expenses. Expenses consist of the costs for snacks and refreshments, spotlight rental, and materials for the costumes and scenery.

Questions

1. The committee decides to charge the same admission fee for students and adults. The table shows the attendance for each night of the play. Write an expression to find the admission income. Let f represent the admission fee. Then evaluate the expression when the admission fee is $4.

Night	Attendance
Friday	216
Saturday	268
Sunday	300

2. Suppose the committee charges admission fees of $5 for adults and $3 for students. The admission profits are shown in the table. Solve the equation $5a + 3(216 - a) = 888$ to find the number of adults a and the number of students that attended the play Friday night.

Night	Admission Profits
Friday	$888
Saturday	$1084
Sunday	$1240

3. You make a list of expenses for the play. The school has $1000 budgeted for expenses. Solve the inequality $3(230 + s) + 80 + 110 \leq 1000$ to determine how much money you have remaining to spend on spotlight rentals for each night.

Expense	Cost
Snacks/refreshments	$230/night
Spotlight rentals	s/night
Costume materials	$80
Scenery materials	$110

4. Find the school's profit from the play. Use the table in Exercise 2 and assume the entire budget is spent.

5. Suppose the committee charges $3 for adults and $1.50 for students. Use the table from Exercise 1 and the table shown to find the number of adults and students that attended the play each night.

Night	Admission Profits
Friday	$511.50
Saturday	$603
Sunday	$694.50

6. Using the same expenses, compare the profits for both sets of admission prices. What conclusions can you make? Explain in detail any changes you might make.

7. Choose a partner and organize your own play. Set your own admission prices and use the Internet to research costs for expenses. Set up equations for your partner to solve to find the number of adults and students that attended your play. Compare profits from both plays.

Teacher's Notes for School Play Project

For use after Chapter 3

Project Goals

- Add, subtract, and multiply integers.
- Simplify and evaluate expressions.
- Solve equations and inequalities using the distributive property.
- Write and solve equations using addition, subtraction, and division.
- Write and solve equations involving decimals.

Managing the Project

Guiding Students' Work Encourage students to show all of their work when solving the exercises. Question 5 may be a little difficult to set up the equations for Friday, Saturday, and Sunday, but explain that these are modeled after the equation in Question 2, which should make writing the equations clearer.

When organizing their play, remind students to think economically about the expenses and to maximize the profit. You can also encourage students to possibly include new expenses or new ways to earn income at the play. An example of a new expense might be to hire a disc jockey for music for the play.

Rubric for Project

The following rubric can be used to assess student work.

4 All of the expressions are correct. The student has written the correct equations and solved them correctly. The student has compared the profits and made insightful conclusions. If changes were made, the changes are reasonable and clearly stated. The student organizes his or her own play with reasonable admission prices, expenses and profit. The equations are then written correctly and solved. The student's work is neat.

3 All of the expressions are correct. The student has written the correct equations and solved them correctly, but some steps may be missing. The student has correctly compared the profits but the conclusions are a little unclear. The student organizes his or her own play and sets up the equations, but there are some minor errors. The student's work is neat.

2 The student's expressions may have minor errors. The equations are written but not solved correctly in some cases. The student has compared the profits but no conclusions are given. The student organizes his or her own play and sets up the equations, but there are some errors. The student's work is sloppy or incomplete.

1 The student's work has several mathematical errors. The equations are incorrect, making the solutions invalid. The student does not correctly calculate the profits and compare them. No attempt is made to organize his or her own play. The student's work is incomplete or sloppy.

Name ______________________ Date ________

Cooperative Project: Telescope

For use after Chapter 3

Objective **Determine how long you need to save money to buy a telescope.**

Materials index cards, markers or pens, paper, pencil

Investigation *Getting Going* (Before you begin, the teacher will instruct you on how to make the cards used in this game.) This project is for 2 students. You and your partner are saving money to buy a telescope. You already have some money saved and plan on saving a certain amount each week until you can buy the telescope.

Make index cards similar to the ones shown. Then divide the cards into the three piles.

1st Pile		2nd Pile	3rd Pile
\$5	\$30	\$2/week	\$180
\$10	\$40	\$3/week	\$200
\$20	\$50	\$4/week	\$210
\$25	\$60	\$5/week	\$230

- 1st Pile (2 cards): You and your partner each choose a card from the first pile. These cards represent the total amount of money you and your partner have already saved.
- 2nd Pile (2 cards): You and your partner each choose a card from the second pile. These cards represent the amount of money you and your partner are going to save each week.
- 3rd Pile (1 card): You or your partner choose one card from the third pile. This card represents the price of the telescope you and your partner are going to buy.

Questions

1. Write a variable expression for the amount of money you have saved after w weeks. Evaluate the expression when $w = 1, 2, 3,$ and 4.

2. Write a variable expression for the amount of money your partner has saved after w weeks. Evaluate the expression when $w = 1, 2, 3,$ and 4.

3. Write and solve an equation to find the number of weeks it takes for you to save enough money to buy the telescope. Write and solve an equation to find the number of weeks it takes your partner to save enough money to buy the telescope. If necessary, round your answers to the nearest tenth.

4. If possible, write and solve an equation to find the number of weeks until you and your partner have saved the same amount. If necessary, round your answer to the nearest tenth.

5. Choose new numbers from each pile. Then repeat Questions 1–4.

Teacher's Notes for Telescope Project

For use after Chapter 3

Project Goals

- Write and evaluate variable expressions.
- Solve two-step equations.
- Solve equations by combining like terms.
- Solve equations with variables on both sides.

Managing the Project

Classroom Management You may need to instruct the students on how to make the cards. If the index cards are not available use pieces of paper that are roughly equal in size.

You can have students repeat this project several times. Keep in mind that you should have students with the same level of math skills paired together.

Encourage students to make a table of values for Questions 1 and 2.

In some cases, Question 4 does not yield an answer depending on the numbers that students select. For example, if one student selects $60 and $4/week and the other student selects $40 and $2/week, they will never have saved the same amount.

Rubric for Project

The following rubric can be used to assess student work.

4 The students correctly write and evaluate the expressions. The students repeat the project several times. They correctly write and solve the equations each time. All their work is shown. All of their calculations are correct. The students' work is neat.

3 The students correctly write and evaluate the expressions. The students repeat the project several times. They write and solve the equations each time with some minor errors. Not all of their work is shown. Most of their calculations are correct. The students' work is neat.

2 The students write and evaluate the expressions with some minor errors. The students repeat the project more than once. They write and solve the equations with some difficulty. Not all of their work is shown. Some of their calculations are incorrect. The students' work is a little sloppy.

1 The students do not write and evaluate the expressions correctly or do not select the numbers correctly. The students have difficulty doing the project. They incorrectly write the equations. Most of their calculations are incorrect. The students' work is incomplete or sloppy.

CHAPTER 3

Name ______________________________ Date __________

Independent Extra Credit Project: Population

For use after Chapter 3

Objective **Study the populations of certain counties in the United States.**

Materials paper, pencil, access to the Internet

Investigation *Getting Going* The counties described below are four of the top 100 fastest growing counties in the United States.

- Bastrop County, Texas has a population of about 61,500. During the next year, the population will increase by about 2500.
- St. Croix County, Minnesota has a population of about 65,700. During the next year, the population will increase by about 2400.
- Sumter County, Florida has a population of about 54,700. During the next year, the population will increase by about 2800.
- Walton County, Georgia has a population of about 64,400. During the next year, the population will increase by about 2600.

Questions

Assume that the annual population increase for each county remains the same each year to answer the following questions. If necessary, round your answer to the nearest tenth.

1. Write an equation that represents the population P of

a. Bastrop County after t years. **b.** St. Croix County after t years.

c. Sumter County after t years. **d.** Walton County after t years.

2. Complete a table of values for the population of each county for the next 5 years.

3. Write and solve an equation to determine after how many years the populations of the given counties will be the same. Write your answer as a decimal.

a. Bastrop County and Sumter County

b. St. Croix County and Bastrop County

c. Walton County and St. Croix County

d. St. Croix County and Sumter County

4. Write an expression that represents the total population of the four counties after t years.

5. Find the population of the county where you live and your bordering counties. Are the populations of these counties increasing or decreasing? Estimate the increase or decrease in your county's population each year. Predict the population of your county after 5 years.

CHAPTER 3 Continued

Teacher's Notes for Population Project

For use after Chapter 3

Project Goals

- Write variable expressions.
- Evaluate variable expressions.
- Simplify variable expressions.
- Solve equations using addition, subtraction, and division.
- Write and solve equations with variables on both sides.

Managing the Project

Guiding Students' Work You can direct students to the link *http://eire.census.gov/popest/data/counties/tables.php* for help with finding county populations. Students should choose the latest annual population estimate by county table from the list of popular tables.

Rubric for Project

The following rubric can be used to assess student work.

4 All of the expressions are written correctly. The tables are neat and correct. The student correctly writes and solves all the equations. The student's work is neat.

3 All of the expressions are written correctly. The tables are correct. The student writes and solves the equations with some minor errors. The student's work is neat.

2 Most of the expressions are written correctly. The tables are correct, but are a little sloppy. The student writes and solves the equations with a few errors. The student's work is a little sloppy.

1 The expressions are written with some errors. The tables are incorrect or incomplete. The equations are written incorrectly. The student's work is incomplete or sloppy.

CHAPTER **3** Name ____________________ Date ________

Cumulative Practice

For use after Chapter 3

Write the product using an exponent. (Lesson 1.2)

1. $5 \cdot 5 \cdot 5$ **2.** $24 \cdot 24$ **3.** $m \cdot m \cdot m \cdot m \cdot m$

Evaluate the expression when $a = 3$ and $b = 5$. (Lesson 1.3)

4. $8a - 3b$ **5.** $4(a + b)^2$ **6.** $\frac{2a^2}{6b + 6}$

Evaluate the expression when $x = 7$ and $y = -3$. (Lesson 1.4)

7. $-|y| + 16$ **8.** $2x + (-y)$

9. $4|-y| - 3|x|$ **10.** $-3|x|$

Simplify. (Lessons 1.5–1.7)

11. $-2 + 6$ **12.** $1 + (-5)$ **13.** $-4 - (-13)$

14. $-16 + (-3)$ **15.** $10 - (-7)$ **16.** $-20 + 9$

17. $-21 - 11$ **18.** $-1(-15)$ **19.** $30 \div (-2)$

20. $3(-60)$ **21.** $\frac{-33}{3}$ **22.** $-5(-20)$

Identify the property that the statement illustrates. (Lesson 2.1)

23. $-3 + 8 = 8 + (-3)$ **24.** $(9 \cdot 4^3) \cdot 2 = 9 \cdot (4^3 \cdot 2)$

25. $-5b \cdot 1 = -5b$ **26.** $k + 0 = k$

Use the distributive property to write an equivalent variable expression. (Lesson 2.2)

27. $-9(x - 4)$ **28.** $(6 + t)(-7)$ **29.** $5(3y - 8)$ **30.** $(2 + 4z)12$

Write the verbal statement as an equation. Then solve the equation. (Lessons 2.4–2.6)

31. The sum of -5 and b is 15. **32.** The quotient of p and 12 is -9.

33. The product of -6 and x is -96. **34.** The sum of $-a$ and -17 is -31.

35. The quotient of y and 3 is -4. **36.** The product of x and -3 is 90.

37. You are going on a school trip to the Grand Canyon. The trip costs \$1100 per student. Through your savings and fundraisers you have raised \$800 for the trip. The last fundraiser is a sandwich sale. For each sandwich you sell, you make a profit of \$3.75. How many sandwiches do you need to sell to pay for the rest of your trip? (Lessons 3.1–3.3)

Name ______________________________ Date __________

Cumulative Practice

For use after Chapter 3

Solve the equation. (Lessons 3.1–3.3)

38. $19 + z = 36$

39. $8k + 15 = -57$

40. $\frac{d}{4} - 12 = -23$

41. $29n + 14 - 17n = 62$

42. $18 = 2(f - 3)$

43. $48 - (5x + 7) = -9$

44. $5m + 31 = -53 - 9m$

45. $7s + 32 = 11s$

46. $13 + 12x = -6x + 49$

47. $-3t + 11 = 4t - 17$

48. $9(3y - 9) = 32y + 19$

49. $25x + 3 = 2(10x - 36)$

Solve the inequality. Graph your solution. (Lessons 3.4–3.6)

50. $p + 5 > 2$

51. $m - 6 \leq 14$

52. $9 \geq y - 3$

53. $g - 17 < -9$

54. $\frac{x}{2} \leq -7$

55. $-13n > 78$

56. $\frac{y}{-3} < 8$

57. $4t \geq -60$

58. $3x + 5 \leq 23$

59. $15 - 3s < -6$

60. $22p - 14 > 13p + 31$

61. $-8 - 5w \geq w - 74$

Answers

Lesson 3.1

Practice A

1. Yes **2.** No **3.** Yes **4.** No **5.** Yes **6.** Yes **7.** 4 **8.** 3 **9.** 0 **10.** 2 **11.** 1 **12.** 3 **13.** 16 **14.** 12 **15.** 50 **16.** 8 **17.** 9 **18.** 30 **19.** -1 **20.** -6 **21.** 5 **22.** 7 **23.** A **24.** \$50 **25.** C **26.** \$14

27. a.

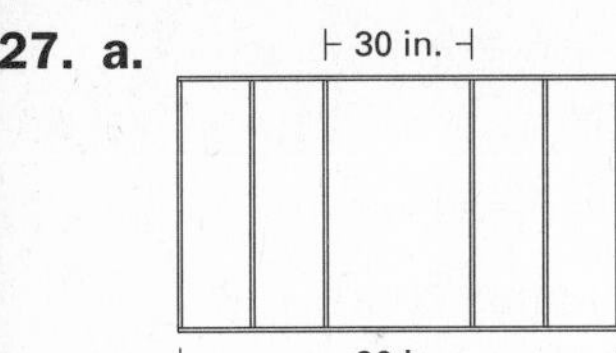

b.

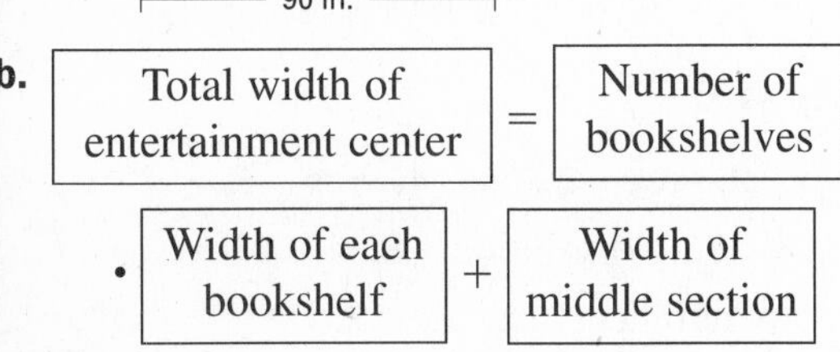

c. $90 = 4w + 30$ **d.** $w = 15$ in.

Practice B

1. Yes **2.** Yes **3.** No **4.** No **5.** Yes **6.** No **7.** 2 **8.** -7 **9.** -3 **10.** 1 **11.** 11 **12.** 10 **13.** -17 **14.** -18 **15.** -24 **16.** 60 **17.** 126 **18.** -75 **19.** -42 **20.** 99 **21.** 135 **22.** $14 - 3x = 26; x = -4$

23. $-7 - 5x = 28; x = -7$

24. $11 - \frac{x}{8} = 15; x = -32$

25. $-16 + \frac{x}{2} = 35; x = 102$

26. $39 - x = -19; x = 58$ **27.** 5 months

28. 5 compact discs **29.** \$8 **30.** 89 students

31. a. 14 min **b.** 20 min

Practice C

1. 6 **2.** -48 **3.** -6 **4.** 8 **5.** -12 **6.** 3 **7.** 17.2 **8.** 27.5 **9.** 12 **10.** 21 **11.** 8.4 **12.** -23 **13.** -18 **14.** 12 **15.** -5

16. $2.5x - 4.7 = 10.3; x = 6$

17. $8.3x + 5.6 = -27.6; x = -4$

18. $\frac{x}{1.8} - 9.4 = 5.6; x = 27$

19. $7.1 + \frac{x}{0.9} = 16.1; x = 8.1$

20. a. $170 + 30m$

m	1	2	3
Amount saved	\$200	\$230	\$260

m	4	5	6
Amount saved	\$290	\$320	\$350

b.

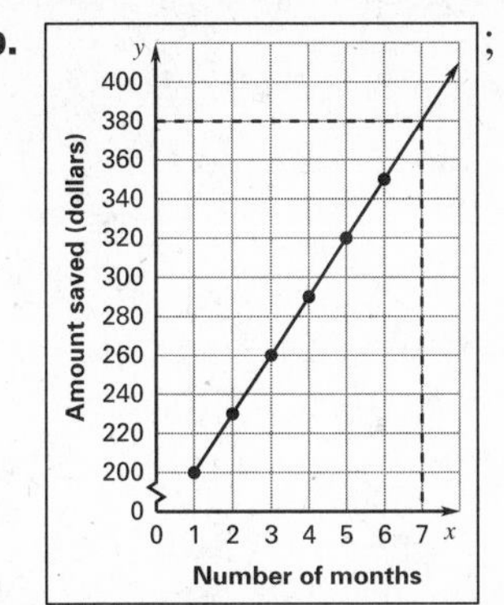

; 7 months

c. $170 + 30m = 380; m = 7$

21. a. $6.4x - 9.2 = 42, 5.3y + 4.9 = 42$

b. $x = 8, y = 7$

Study Guide

1. 5 **2.** 6 **3.** 15 **4.** 20 **5.** -140 **6.** 63 **7.** 96 **8.** 90 **9.** -3 **10.** 3 **11.** 28 **12.** -99

Real-World Problem Solving

1. \$88 **2.** $0.06p$

3. Extra costs = Sales tax + Total fees

4. $C = 0.06p + 88$ **5.** \$1034.80 **6.** \$19,250

Challenge Practice

1. 1.5 **2.** 6.5 **3.** 31.5 **4.** 9.02 **5.** 6 CDs **6.** 7 paperback books **7.** $18 - \frac{a}{4} = 54; -144$

8. $25 - \frac{a}{8} = 20; 40$ **9.** 39; Because $\frac{35}{5} = 7$ and $\frac{x-4}{5} = 7, x - 4 = 35$. So, $x = 39$. The solution is correct because $\frac{39-4}{5} = \frac{35}{5} = 7$.

Lesson 3.2

Activity Master

1. 1 **2.** 2 **3.** 3 **4.** 4 **5.** 4 **6.** 2

Lesson 3.2 *continued*

7. Write the original equation, $x + 5 + 2x = 11$. Then group and combine like terms. Subtract 5 from each side of the equation. Then divide each side of the equation by 3.

Practice A

1. No **2.** Yes **3.** Yes **4.** No **5.** 5 **6.** 1 **7.** -2 **8.** 12 **9.** -3 **10.** -15 **11.** 0 **12.** -11 **13.** -5 **14.** -2 **15.** 9 **16.** 7 **17.** 7 **18.** 5 **19.** 3 **20.** 2 **21.** 6 in. × 9 in. **22.** $8 **23.** $12

Practice B

1. 5 **2.** -8 **3.** -1 **4.** 9 **5.** -7 **6.** -11 **7.** -6 **8.** -3 **9.** -5 **10.** 8 **11.** 3 **12.** -4 **13.** 10 **14.** 4 **15.** 11 **16.** 8

17. a.

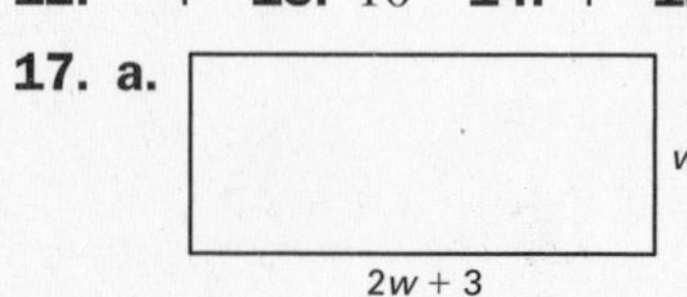

b. $2w + 2(2w + 3) = 48$

c. Width = 7 in., length = 17 in.

18. $485 = 230 + 5(44 + g)$; 7 guests

19. $3(30) + 42h + 195 = 390$; 2.5 hours

Practice C

1. -4 **2.** -15 **3.** 3 **4.** -9 **5.** 4 **6.** 1 **7.** 8 **8.** -5 **9.** 6 **10.** -2 **11.** 12 **12.** 10 **13.** 6.2 **14.** 6 **15.** 12 **16.** 4.9

17. $p + (p + 1) + (p + 1 + 1) = 45$; Kevin has 14 sponsors, Mika has 15 sponsors, and Cheryl has 16 sponsors. **18.** $10.75

19. $2x + 2(2x + 2) = 178$; 58 and 60

20. $(2x - 1) + (2x - 1 + 2) = 104$; 51 and 53

Study Guide

1. 2 pounds **2.** 5 **3.** 5 **4.** 11 **5.** 3 **6.** 1 **7.** 12

Challenge Practice

1. 12 **2.** $\frac{3}{4}$ **3.** 1 **4.** No solution **5.** 12 fish **6.** 6, 11, and 22 **7.** 5 and 19 **8.** 2 **9.** 3

Lesson 3.3

Practice A

1. Yes **2.** No **3.** No **4.** Yes **5.** 11 **6.** 1 **7.** -7 **8.** 2 **9.** No solution **10.** 4 **11.** 6 **12.** 3 **13.** -8 **14.** Every number **15.** 8 **16.** 5 **17.** $5 - 6x = -11 + 2x$; $x = 2$

18. $-7x - 4 = 13 - 6x$; $x = -17$

19. $8x + 5 = 5x - 13$; $x = -6$

20. $10x - 1 = -2x + 35$; $x = 3$ **21.** 6 **22.** 5

23. $60 + 5w = 45 + 8w$; 5 weeks

24. $4t + 48 = t + 282$; tennis court is 78 feet long, football field is 360 feet long

Practice B

1. No **2.** Yes **3.** No **4.** Yes **5.** -7 **6.** -18 **7.** No solution **8.** 6 **9.** -5 **10.** Every number **11.** 2 **12.** 94 **13.** No solution **14.** 0 **15.** -6 **16.** 17

17. $-13x + 20 = -11x + 38$; $x = -9$

18. $6x - 17 = 47 + 10x$; $x = -16$

19. $-10x - 29 = -18x + 91$; $x = 15$

20. $17x - 56 = 10x - 63$; $x = -1$

21. 309 units **22.** 54 units **23.** $3.50; $224 **24.** 8 years

Practice C

1. 4 **2.** -11 **3.** -17 **4.** 75 **5.** -56 **6.** 1 **7.** -1 **8.** 24 **9.** 6 **10.** 23.47

11. $6 + 0.25x = 0.8x - 5$; $x = 20$

12. $12 - 1.8x = 22.8 - 2.7x$; $x = 12$

13. $-20 - 4.6x = -2.3x + 16.8$; $x = -16$

14. $7x - 10.75 = 6.8x + 0.95$; $x = 58.5$

15. 192.3 units **16.** 68 units **17.** 12.48 units **18.** 29.42 units

19. $1450 - 105.75m = 825 + 144.25m$; 2.5 months **20.** $12 + 2x = 4x$; 6 feet

Study Guide

1. 2 **2.** -14 **3.** 3 **4.** 125 minutes **5.** Every number **6.** No solution **7.** 152 units

Lesson 3.3 *continued*

Challenge Practice

1. 3 **2.** −1 **3.** 4 **4.** −6
5. about 1538 copies **6.** 40 smoothies
7. −32 **8.** 3

Lesson 3.4

Technology Activity

1. $x < -15$;
−18 −17 −16 −15 −14 −13 −12

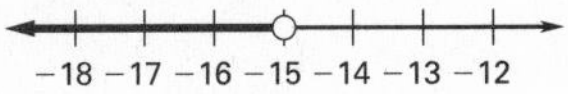

2. $x \le -39$;
−42 −41 −40 −39 −38 −37 −36

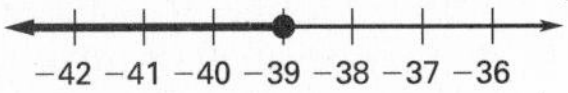

3. $x > 1$;
−2 −1 0 1 2 3 4

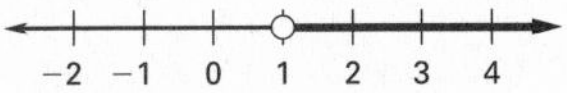

4. $x \ge 43$;
40 41 42 43 44 45 46

5. $x \ge -4.93$;
−4.95 −4.93 −4.91

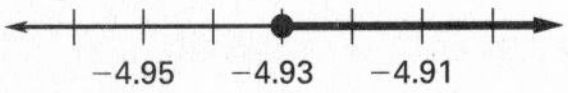

6. $x < 19.3$;
19.1 19.3 19.5

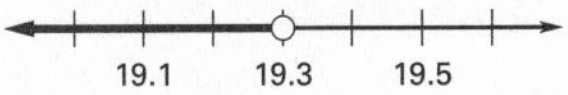

7. $x \le 39.9$;
39.7 39.9 40.1

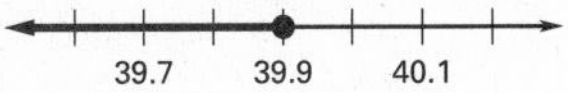

8. $x < 14.62$;
14.60 14.62 14.64

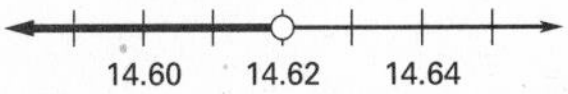

Practice A

1. Yes **2.** No **3.** Yes **4.** No **5.** $x \ge 24$
6. $x \le 325$ **7.** $x \le 180$ **8.** $x \ge 580$
9. $x > -5$ **10.** $x \le 8$ **11.** $x \ge -11$
12. $x < -40$ **13.** $x \le 16$ **14.** $x > 75$

15. $x > 7$;
2 3 4 5 6 7 8 9 10

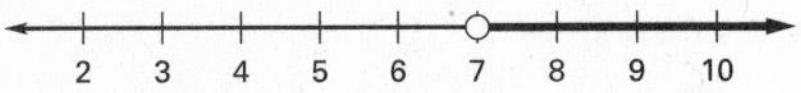

16. $x \le 17$;
12 13 14 15 16 17 18 19 20

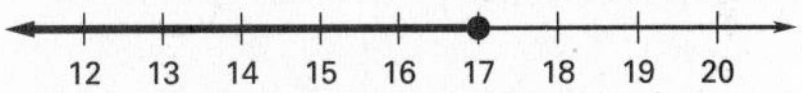

17. $-22 \ge x$;
−28 −27 −26 −25 −24 −23 −22 −21 −20

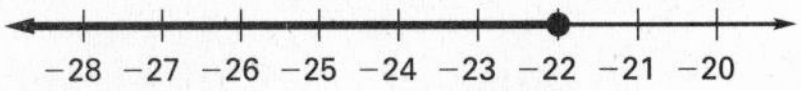

18. $9 < x$;
5 6 7 8 9 10 11 12 13

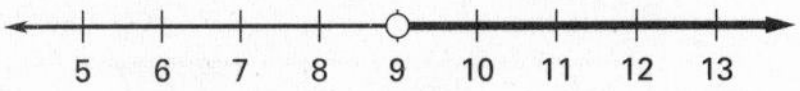

19. $x < 53$;
48 49 50 51 52 53 54 55 56

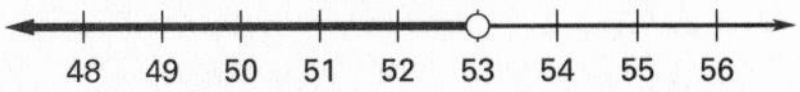

20. $x \ge -3$;
−4 −3 −2 −1 0 1 2 3 4

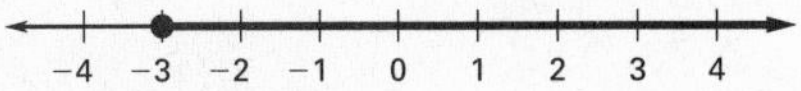

21. $x \le -13$;
−18 −17 −16 −15 −14 −13 −12 −11 −10

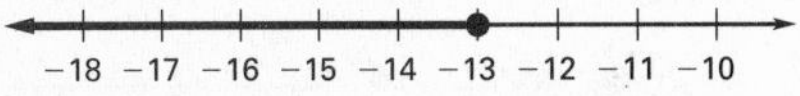

22. $x > 0$;
−4 −3 −2 −1 0 1 2 3 4

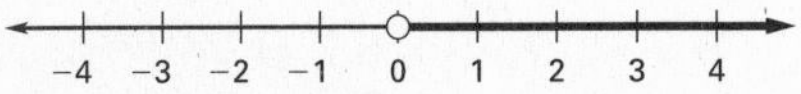

23. $x < -14$;
−18 −17 −16 −15 −14 −13 −12 −11 −10

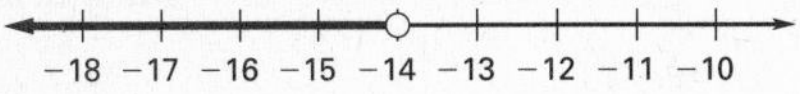

24. $x \ge 13$;
12 13 14 15 16 17 18 19 20

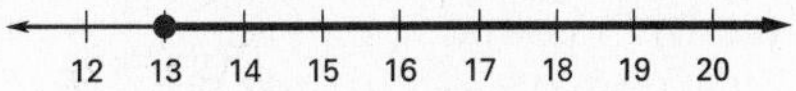

25. $6 \le x$;
2 3 4 5 6 7 8 9 10

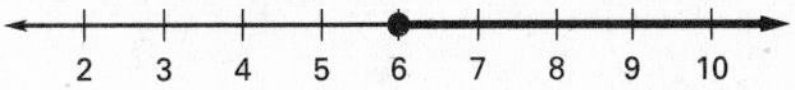

26. $-16 > x$;
−18 −17 −16 −15 −14 −13 −12 −11 −10

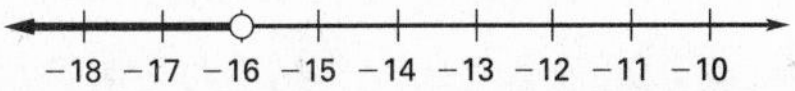

27. $x \ge 11.6$;
11.0 11.1 11.2 11.3 11.4 11.5 11.6 11.7 11.8

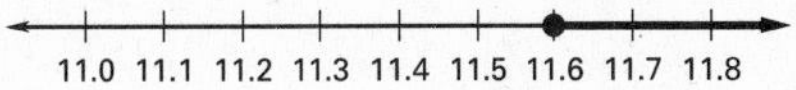

28. $x \le 3.2$;
2.7 2.8 2.9 3.0 3.1 3.2 3.3 3.4 3.5

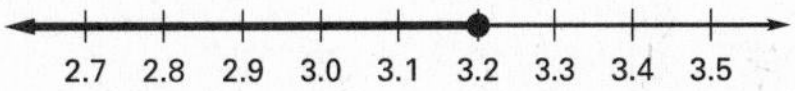

29. $x < 19$;
12 13 14 15 16 17 18 19 20

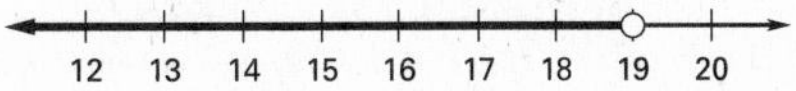

30. $1.6 > x$;
1.1 1.2 1.3 1.4 1.5 1.6 1.7 1.8 1.9

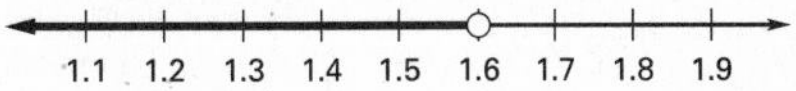

31. $x + 78 \ge 120$; $x \ge 42$

Practice B

1. Yes **2.** No **3.** Yes **4.** No
5. $9.4 + x \le 14.1$ **6.** $32 + x - 18 > -3$
7. $0.6 + 4.7 + x \ge -5.6$ **8.** $x - 6.88 < 22.74$
9. C **10.** A **11.** B **12.** D

13. $x < -19$;
−25 −24 −23 −22 −21 −20 −19 −18 −17

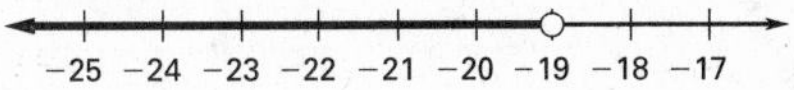

14. $x \ge 8$;
2 3 4 5 6 7 8 9 10

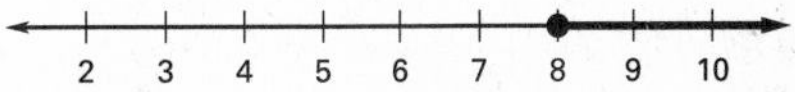

15. $x \le 32$;
22 24 26 28 30 32 34 36 38

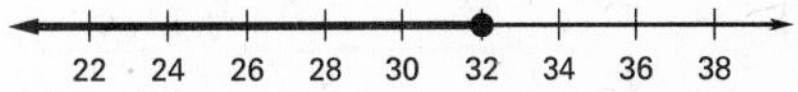

16. $x > 7$;
5 6 7 8 9 10 11 12 13

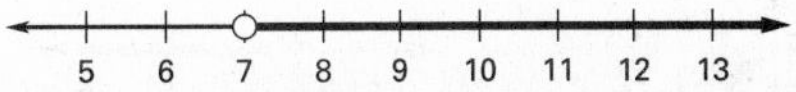

17. $-30 \le x$;
−32 −31 −30 −29 −28 −27 −26 −25 −24

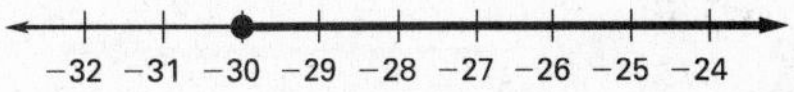

18. $x > -4.4$;
−4.6 −4.5 −4.4 −4.3 −4.2 −4.1 −4.0 −3.9 −3.8

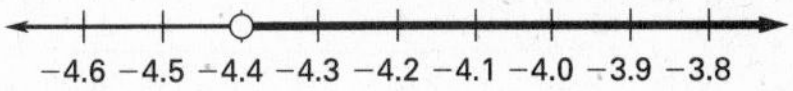

19. $x < -8$;
−11 −10 −9 −8 −7 −6 −5 −4 −3

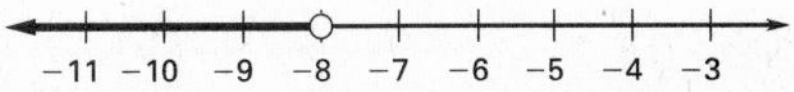

20. $x > 0$;
−4 −3 −2 −1 0 1 2 3 4

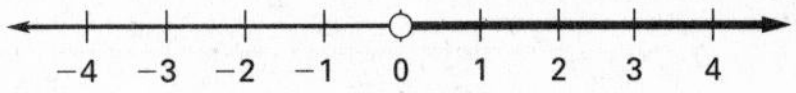

21. $1.36 \le x$;
1.33 1.34 1.35 1.36 1.37 1.38 1.39

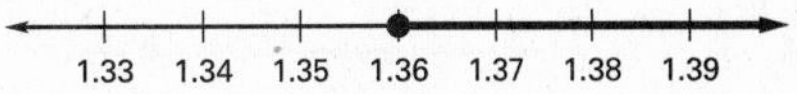

22. $x \le -26.38$
−26.40 −26.39 −26.38 −26.37 −26.36

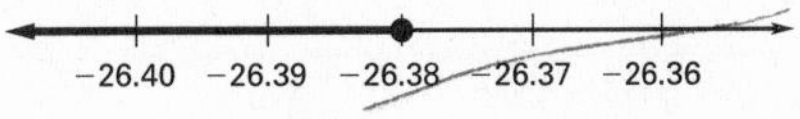

23. $8.71 \ge x$;
8.69 8.70 8.71 8.72 8.73

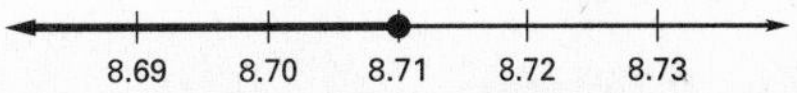

24. $-114.34 < x$
−114.35 −114.34 −114.33 −114.32

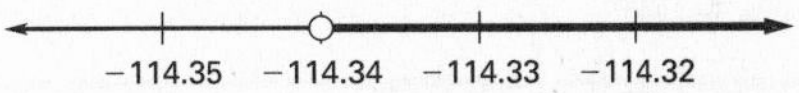

25. Friday: $t + 3488 \le 5400$; $t \le 1912$;
Saturday: $t + 4109 \le 5400$; $t \le 1291$;
Sunday: $t + 4573 \le 5400$; $t \le 827$

Lesson 3.4 *continued*

26. a. 1825 pounds
b. $w + 1825 \le 2000$; $w \le 175$

Practice C

1. $3.5 + x \ge -6.7$ **2.** $12.3 + x - 5.6 < -2.8$
3. $-14.3 \le -9.1 + x$
4. $17.07 > 31.02 + x - 27.64$
5. C **6.** A **7.** D **8.** B
9. $x < -4.8$;
−5.2 −5.1 −5.0 −4.9 −4.8 −4.7 −4.6 −4.5 −4.4
10. $x \ge -23.4$;
−23.7 −23.6 −23.5 −23.4 −23.3 −23.2 −23.1
11. $1.2 > x$;
0.8 0.9 1.0 1.1 1.2 1.3 1.4 1.5 1.6
12. $3.6 \le x$;
3.2 3.3 3.4 3.5 3.6 3.7 3.8 3.9 4.0
13. $x < 3.2$;
2.8 2.9 3.0 3.1 3.2 3.3 3.4 3.5 3.6
14. $x \le 9.5$;
9.1 9.2 9.3 9.4 9.5 9.6 9.7 9.8 9.9
15. $0.9 \ge x$;
0.5 0.6 0.7 0.8 0.9 1.0 1.1 1.2 1.3
16. $6.5 > x$;
6.1 6.2 6.3 6.4 6.5 6.6 6.7 6.8 6.9
17. $x > -5$;
−9 −8 −7 −6 −5 −4 −3 −2 −1
18. $x < 10$;
6 7 8 9 10 11 12 13 14
19. $x > -1.25$;
−1.75 −1.50 −1.25 −1.00 −0.75
20. $x \le 38.4$;
38.0 38.1 38.2 38.3 38.4 38.5 38.6 38.7 38.8
21. $8 \le x$;
2 3 4 5 6 7 8 9 10
22. $x \ge 34.2$;
33.9 34.0 34.1 34.2 34.3 34.4 34.5
23. $-0.5 > x$;
−0.8 −0.7 −0.6 −0.5 −0.4 −0.3 −0.2
24. $0.6 \ge x$;
0.3 0.4 0.5 0.6 0.7 0.8 0.9
25. $5.5 + 5.75 + 7 + 6.5 + 5.625 + w > 33.5$;
$w > 3.125$;
3.123 3.124 3.125 3.126 3.127

Study Guide

1. $p \ge 85$ **2.** $c \le 7500$
3. $y < -1$;
−7 −6 −5 −4 −3 −2 −1 0 1 2 3
4. $t > -8$;
−13 −12 −11 −10 −9 −8 −7 −6 −5 −4 −3

5 $m \le -16$;
−30 −28 −26 −24 −22 −20 −18 −16 −14 −12 −10
6. $x \ge 3$;
−2 −1 0 1 2 3 4 5 6 7 8
7. $x \ge \$37$

Real-World Problem Solving

1. $600 + x \ge 700$ **2.** $x \ge 100$
3. $600 + x \le 2800$ **4.** $x \le 2200$

5. Yes; the inequality in Exercise 2 shows that she needs 100 or more micrograms of vitamin A, and a 750 microgram capsule gives her that.

6. Yes; no; the inequality in Exercise 4 shows that Trista should take at most 2200 micrograms. There are 1500 micrograms in 2 capsules and 2250 micrograms in 3 capsules, so with 3 capsules her daily intake is too great.

Challenge Practice

1. $t \le -1.1$;
−1.1
−4.0 −3.0 −2.0 −1.0 0.0 1.0
2. $r > 24$;
19 20 21 22 23 24 25 26 27 28 29
3. $x > 2.18$;
2.18
1.60 1.80 2.00 2.20 2.40 2.60
4. $t \ge -4$;
−8 −7 −6 −5 −4 −3 −2 −1 0 1 2
5. $8 + n < -14$; $n < -22$ **6.** $14 \ge n - 4$; $n \le 18$
7. $x < -78.5$;
−78.5
−84 −82 −80 −78 −76 −74
8. $215.5 < w + 50.5$; $w > 165$
9. r is less than or equal to 3.6 and greater than −4.3.
−4.3 3.6
−5 −4 −3 −2 −1 0 1 2 3 4 5

Lesson 3.5

Practice A

1. No **2.** Yes **3.** No **4.** Yes
5. C **6.** A **7.** B **8.** D
9. $x > -6$;
−8 −7 −6 −5 −4 −3

.esson 3.5 *continued*

0. $x \le 64$; 61 62 63 64 65 66

1. $x \ge -7$; −8 −7 −6 −5 −4 −3

2. $x < 3$; −1 0 1 2 3 4

3. $x > -17$; −19 −18 −17 −16 −15 −14

4. $x \ge 11$; 9 10 11 12 13 14

5. $x < 54$; 51 52 53 54 55 56

6. $x \le 121$; 118 119 120 121 122 123

7. $x \ge -10$; −12 −11 −10 −9 −8 −7

8. $x > -60$; −62 −61 −60 −59 −58 −57

9. $x > -6$; −8 −7 −6 −5 −4 −3

0. $x \ge 9$; 9 10 11 12 13 14

1. $x \le -8$; −12 −11 −10 −9 −8 −7

2. $x < 5$; 0 1 2 3 4 5

3. $x < 14$; 9 10 11 12 13 14

4. $x \ge 52$; 51 52 53 54 55 56

5. $15d < 75$; $d < 5$, less than 5 DVDs

6. $3t \ge 1575$; $t \ge 525$, at least 525 tickets

7. $2.5d \ge 30$; $d \ge 12$, at least 12 deliveries

ʼractice B

1. Yes **2.** No **3.** Yes **4.** No **5.** B **6.** A

7. C **8.** D

9. $x \ge -220$; −222 −221 −220 −219 −218 −217

0. $x > -750$; −752 −751 −750 −749 −748 −747

1. $x > 16$; 14 15 16 17 18 19

2. $x \le -19$; −22 −21 −20 −19 −18 −17

3. $x \le -7$; −10 −9 −8 −7 −6 −5

4. $x < -4$; −7 −6 −5 −4 −3 −2

5. $42.8 < x$; 42.6 42.7 42.8 42.9 43.0 43.1

16. $-41.5 \ge x$

−43.0 −42.5 −42.0 −41.5 −41.0 −40.5

17. $x \ge 9.23$; 9.21 9.22 9.23 9.24 9.25 9.26

18. $x > -15.68$

−15.70 −15.69 −15.68 −15.67 −15.66 −15.65

19. $x \le -7.5$; −9.0 −8.5 −8.0 −7.5 −7.0 −6.5

20. $x > 50$; 48 49 50 51 52 53

21. $x \ge 25$; 23 24 25 26 27 28

22. $x \le -4.5$; −6.0 −5.5 −5.0 −4.5 −4.0 −3.5

23. $x \le 27.94$; 27.92 27.93 27.94 27.95 27.96 27.97

24. $x < -85.89$

−85.92 −85.91 −85.90 −85.89 −85.88 −85.87

25. $\frac{x}{3.5} \ge 7.8$; $x \ge 27.3$ **26.** $-5x < -1.6$; $x > 0.32$ **27.** $-0.9x > 27$; $x < -30$

28. $\frac{x}{-4.75} \ge -20$; $x \le 95$ **29.** $x \ge 75$

30. $4.5p \ge 7200$; $p \ge 1600$, at least 1600 passes

Practice C

1. C **2.** D **3.** B **4.** A

5. $x \ge -7.29$; −7.31 −7.30 −7.29 −7.28 −7.27 −7.26

6. $x < 34.04$; 34.01 34.02 34.03 34.04 34.05 34.06

7. $x > 56$; 54 55 56 57 58 59

8. $x \ge 16$; 14 15 16 17 18 19

9. $x \le -8$; −10 −9 −8 −7 −6 −5

10. $x < 64.8$; 64.5 64.6 64.7 64.8 64.9 65.0

11. $x < -564$; −567 −566 −565 −564 −563 −562

12. $x > 136.5$; 135.5 136.0 136.5 137.0 137.5 138.0

13. $x < 6.65$; 6.62 6.63 6.64 6.65 6.66 6.67

14. $x \ge -171$; −173 −172 −171 −170 −169 −168

15. $x \le -0.5$; −2.0 −1.5 −1.0 −0.5 0.0 0.5

16. $-0.75 > x$; −1.50 −1.25 −1.00 −0.75 −0.50 −0.25

Lesson 3.5 *continued*

17. $x \geq 42$; 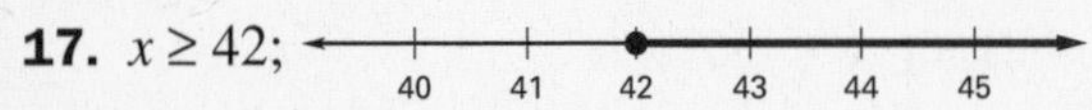

18. $x > 32{,}500$

19. $x \leq -0.37$

20. $x \leq 12$;

21. $2.75\ell < 14.85$; $\ell < 5.4$ cm

22. $3.8\ell \leq 32.87$; $\ell \leq 8.65$ ft

23. $5.625\ell > 67.5$; $\ell > 12$ yd

24–27. Answers will vary.

Study Guide

1. $h < -24$;

2. $u > -7$;

3. $y \leq 9$;

4. $x \geq -54$;

5. $x \geq 12$, at least 12 hours

Challenge Practice

1. $x \geq 5$;

2. $a < 52$;

3. $n > -10.08$

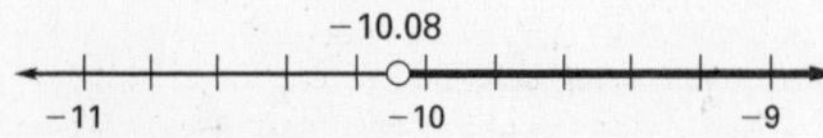

4. $w \geq -137.7$

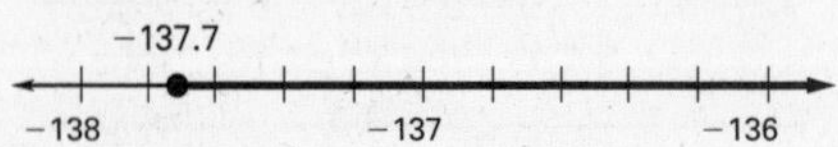

5. $t \geq 50$ mi/h **6.** $c \geq 200$ cards **7.** $w \leq 9$ units

8. $x < 21$ and $x \geq -4$

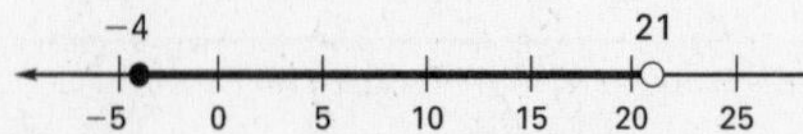

9. $x < 2$;

Lesson 3.6

Practice A

1. No **2.** Yes **3.** No **4.** No **5.** No

6. Yes **7.** Yes **8.** Yes **9.** B **10.** D

11. C **12.** A

13. $x < 0$;

14. $x \geq 3$;

15. $x < 8$;

16. $x \leq -5$;

17. $x > 6$;

18. $x \leq -16$;

19. $x < -21$;

20. $x \geq -36$;

21. $580 \leq x$;

22. $x > -10$;

23. $x \leq 4$;

24. $x > -2$;

25. $x \leq 5$;

26. $x < 17$;

27. $x \geq 7$;

28. $736 + 10p > 1076$; $p > 34$, your team must score, on average, more than 34 points per game.

Practice B

1. No **2.** Yes **3.** Yes **4.** No

5. D **6.** B **7.** A **8.** C

9. $x < -12$;

10. $-216 < x$;

11. $x \leq -3$;

12. $x > 1$;

13. $x < 11$;

Lesson 3.6 *continued*

14. $x \geq -13$; −15 −14 −13 −12 −11 −10

15. $x \geq -74$; −76 −75 −74 −73 −72 −71

16. $x > -12$; −15 −14 −13 −12 −11 −10

17. $x > -5$; −7 −6 −5 −4 −3 −2

18. $x \geq 35$; 33 34 35 36 37 38

19. $x < 25$; 22 23 24 25 26 27

20. $-20 > x$; −40 −30 −20 −10 0 10

21. $x \leq 11$; 8 9 10 11 12 13

22. $x > 7$; 6 7 8 9 10 11

23. $-31 \leq x$; −32 −31 −30 −29 −28 −27

24. $160 + 22r < 8r + 22r$; $20 < r$, you need to play more than 20 rounds of golf.

25. $x > 9$ units **26.** $x > 7$ units

Practice C

1. Yes **2.** Yes **3.** No **4.** No **5.** C **6.** D **7.** A **8.** B

9. $x \geq -13$; −15 −14 −13 −12 −11 −10

10. $x > -128$; −130 −129 −128 −127 −126 −125

11. $x < 39.2$; 38.9 39.0 39.1 39.2 39.3 39.4

12. $x \geq -28$; −30 −29 −28 −27 −26 −25

13. $x > 29$; 27 28 29 30 31 32

14. $x \leq -7.7$; −8.0 −7.9 −7.8 −7.7 −7.6 −7.5

15. $x < 6.25$; 5.50 5.75 6.00 6.25 6.50 6.75

16. $x < -3.4$; −3.7 −3.6 −3.5 −3.4 −3.3 −3.2

17. $26.1 > x$; 25.8 25.9 26.0 26.1 26.2 26.3

18. $x \leq 6$; 3 4 5 6 7 8

19. $x > -70.8$; −71.0 −70.9 −70.8 −70.7 −70.6 −70.5

20. $x \geq 8.7$; 8.5 8.6 8.7 8.8 8.9 9.0

21. $x < 800$; 500 600 700 800 900 1000

22. $x \geq 2.5$; 1.5 2.0 2.5 3.0 3.5 4.0

23. $6.95h + 42.90 > 168$; $h > 18$, more than 18 hours **24.** $2 + 1.75m + 3.5 \leq 16$; $m \leq 6$, you can travel at most 6 miles.

Study Guide

1. $x \geq \$7$ per mile **2.** $x > 1$ hour

3. $x \geq 120$;

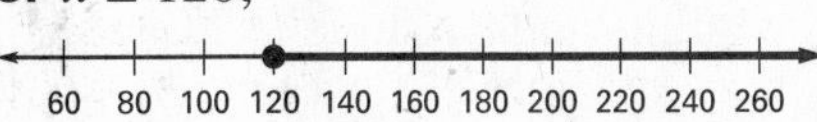

4. $m < 8$; 5 6 7 8 9 10 11 12 13 14 15

5. $x \leq -13$;

−18 −17 −16 −15 −14 −13 −12 −11 −10 −9 −8

6. $x > 33$ times

Challenge Practice

1. $b < -5$; −10 −9 −8 −7 −6 −5 −4 −3 −2 −1 0

2. $x \leq 2$; −3 −2 −1 0 1 2 3 4 5 6 7

3. $x < -20$; −26 −24 −22 −20 −18 −16

4. $p \geq -\frac{22}{3}$;

$-\frac{22}{3}$

−13 −12 −11 −10 −9 −8 −7 −6 −5 −4 −3

5. 6 games **6.** 9 binders **7.** You have to rent the boat for over 2.5 hours. **8.** $x < 1$

9. $\ell \geq 11$ units

Review and Projects

Chapter Review Games and Activities

1 3	2 5	3 7	■	4 1	5 5
■	6 9	0	7 8	■	3
8 2	■	■	9 2	1	■
10 1	0	11 1	8	■	12 3
5	■	9	■	13 3	4
■	14 4	5	9	■	3

Review and Projects *continued*

Real-Life Project

1. $216f + 268f + 300f = 784f$; \$3136

2. 120 adults and 96 students attended the play.

3. You have \$40 to spend on spotlight rentals for each night. **4.** The school's profit is \$2212 for the three-day play.

5. Friday: 125 adults and 91 students attended the play. Saturday: 134 adults and 134 students attended the play. Sunday: 163 adults and 137 students attended the play.

6. The school made a profit of \$2212 when charging \$5 for adults and \$3 for students. The school made a profit of \$809 when charging \$3 for adults and \$1.50 for students. Conclusions will vary. *Sample answer:* The school obviously makes more profit with the higher admission costs. In order to use the second set of admission prices, you could possibly cut back on the expenses by maybe cutting out snacks and refreshments, or not having spotlights. This will increase your profit.

7. Check student's work.

Cooperative Project

1–5. Answers will vary.

Independent Extra Credit Project

1. a. $P = 61{,}500 + 2500t$
b. $P = 65{,}700 + 2400t$ **c.** $P = 54{,}700 + 2800t$
d. $P = 64{,}400 + 2600t$

2.

Bastrop County	
Year	**Population**
1	61,500 + 2500(1) = 64,000
2	61,500 + 2500(2) = 66,500
3	61,500 + 2500(3) = 69,000
4	61,500 + 2500(4) = 71,500
5	61,500 + 2500(5) = 74,000

St. Croix County	
Year	**Population**
1	65,700 + 2400(1) = 68,100
2	65,700 + 2400(2) = 70,500
3	65,700 + 2400(3) = 72,900
4	65,700 + 2400(4) = 75,300
5	65,700 + 2400(5) = 77,700

Sumter County	
Year	**Population**
1	54,700 + 2800(1) = 57,500
2	54,700 + 2800(2) = 60,300
3	54,700 + 2800(3) = 63,100
4	54,700 + 2800(4) = 65,900
5	54,700 + 2800(5) = 68,700

Walton County	
Year	**Population**
1	64,400 + 2600(1) = 67,000
2	64,400 + 2600(2) = 69,600
3	64,400 + 2600(3) = 72,200
4	64,400 + 2600(4) = 74,800
5	64,400 + 2600(5) = 77,400

3. a. 22.7 years **b.** 42 years **c.** 6.5 years **d.** 27.5 years

4. $246{,}300 + 10{,}300t$ **5.** Check student's work.

Cumulative Practice

1. 5^3 **2.** 24^2 **3.** m^5 **4.** 9 **5.** 256 **6.** $\frac{1}{2}$
7. 13 **8.** 17 **9.** −9 **10.** −21 **11.** 4
12. −4 **13.** 9 **14.** −19 **15.** 17 **16.** −11
17. −32 **18.** 15 **19.** −15 **20.** −180
21. −11 **22.** 100
23. Comm. prop. of add.
24. Assoc. prop. of mult.
25. Identity prop. of mult.
26. Identity prop. of add.
27. $-9x + 36$ **28.** $-42 - 7t$ **29.** $15y - 40$
30. $24 + 48z$ **31.** $-5 + b = 15$; 20

Review and Projects *continued*

32. $\frac{p}{12} = -9; -108$ **33.** $-6x = -96; 16$

34. $-a + (-17) = -31; 14$

35. $\frac{y}{3} = -4; -12$ **36.** $-3x = 90; -30$

37. 80 sandwiches **38.** 17 **39.** −9 **40.** −44

41. 4 **42.** 12 **43.** 10 **44.** −6 **45.** 8

46. 2 **47.** 4 **48.** −20 **49.** −15

50. $p > -3$;

51. $m \le 20$;

52. $y \le 12$;

53. $g < 8$;

54. $x \le -14$;

55. $n < -6$;

56. $y > -24$;

57. $t \ge -15$;

58. $x \le 6$;

59. $s > 7$;

60. $p > 5$;

61. $w \le 11$;